Danuta Zakrzewska, Ernestina Menasalvas, and Liliana Byczkowska-Lipinska (Eds.)

Methods and Supporting Technologies for Data Analysis

Studies in Computational Intelligence, Volume 225

Editor-in-Chief
Prof. Janusz Kacprzyk
Systems Research Institute
Polish Academy of Sciences
ul. Newelska 6
01-447 Warsaw
Poland
E-mail: kacprzyk@ibspan.waw.pl

Further volumes of this series can be found on our homepage: springer.com

Vol. 204. Ajith Abraham, Aboul-Ella Hassanien, and André Ponce de Leon F. de Carvalho (Eds.)
Foundations of Computational Intelligence Volume 4, 2009
ISBN 978-3-642-01087-3

Vol. 205. Ajith Abraham, Aboul-Ella Hassanien, and Václav Snášel (Eds.)
Foundations of Computational Intelligence Volume 5, 2009
ISBN 978-3-642-01535-9

Vol. 206. Ajith Abraham, Aboul-Ella Hassanien, André Ponce de Leon F. de Carvalho, and Václav Snášel (Eds.)
Foundations of Computational Intelligence Volume 6, 2009
ISBN 978-3-642-01090-3

Vol. 207. Santo Fortunato, Giuseppe Mangioni, Ronaldo Menezes, and Vincenzo Nicosia (Eds.)
Complex Networks, 2009
ISBN 978-3-642-01205-1

Vol. 208. Roger Lee, Gongzu Hu, and Huaikou Miao (Eds.)
Computer and Information Science 2009, 2009
ISBN 978-3-642-01208-2

Vol. 209. Roger Lee and Naohiro Ishii (Eds.)
Software Engineering, Artificial Intelligence, Networking and Parallel/Distributed Computing, 2009
ISBN 978-3-642-01202-0

Vol. 210. Andrew Lewis, Sanaz Mostaghim, and Marcus Randall (Eds.)
Biologically-Inspired Optimisation Methods, 2009
ISBN 978-3-642-01261-7

Vol. 211. Godfrey C. Onwubolu (Ed.)
Hybrid Self-Organizing Modeling Systems, 2009
ISBN 978-3-642-01529-8

Vol. 212. Viktor M. Kureychik, Sergey P. Malyukov, Vladimir V. Kureychik, and Alexander S. Malyoukov
Genetic Algorithms for Applied CAD Problems, 2009
ISBN 978-3-540-85280-3

Vol. 213. Stefano Cagnoni (Ed.)
Evolutionary Image Analysis and Signal Processing, 2009
ISBN 978-3-642-01635-6

Vol. 214. Been-Chian Chien and Tzung-Pei Hong (Eds.)
Opportunities and Challenges for Next-Generation Applied Intelligence, 2009
ISBN 978-3-540-92813-3

Vol. 215. Habib M. Ammari
Opportunities and Challenges of Connected k-Covered Wireless Sensor Networks, 2009
ISBN 978-3-642-01876-3

Vol. 216. Matthew Taylor
Transfer in Reinforcement Learning Domains, 2009
ISBN 978-3-642-01881-7

Vol. 217. Horio-Nicolai Teodorescu, Junzo Watada, and Lakhmi C. Jain (Eds.)
Intelligent Systems and Technologies, 2009
ISBN 978-3-642-01884-8

Vol. 218. Maria do do Carmo Nicoletti and Lakhmi C. Jain (Eds.)
Computational Intelligence Techniques for Bioprocess Modelling, Supervision and Control, 2009
ISBN 978-3-642-01887-9

Vol. 219. Maja Hadzic, Elizabeth Chang, Pornpit Wongthongtham, and Tharam Dillon
Ontology-Based Multi-Agent Systems, 2009
ISBN 978-3-642-01903-6

Vol. 220. Bettina Berendt, Dunja Mladenic, Marco de de Gemmis, Giovanni Semeraro, Myra Spiliopoulou, Gerd Stumme, Vojtech Svatek, and Filip Zelezny (Eds.)
Knowledge Discovery Enhanced with Semantic and Social Information, 2009
ISBN 978-3-642-01890-9

Vol. 221. Sebastian Schaffert, Klaus Tochtermann, and Tassilo Pellegrini (Eds.)
Networked Knowledge - Networked Media, 2009
ISBN 978-3-642-02183-1

Vol. 222. Elisabeth Rakus-Andersson, Ronald R. Yager, Nikhil Ichalkaranje, and Lakhmi C. Jain (Eds.)
Recent Advances in Decision Making, 2009
ISBN 978-3-642-02186-2

Vol. 223. Zbigniew W. Ras and Agnieszka Dardzinska (Eds.)
Advances in Data Management, 2009
ISBN 978-3-642-02189-3

Vol. 224. Amandeep S. Sidhu and Tharam S. Dillon (Eds.)
Biomedical Data and Applications, 2009
ISBN 978-3-642-02192-3

Vol. 225. Danuta Zakrzewska, Ernestina Menasalvas, and Liliana Byczkowska-Lipinska (Eds.)
Methods and Supporting Technologies for Data Analysis, 2009
ISBN 978-3-642-02195-4

Danuta Zakrzewska, Ernestina Menasalvas,
and Liliana Byczkowska-Lipinska (Eds.)

Methods and Supporting Technologies for Data Analysis

Springer

Danuta Zakrzewska
Institute of Computer Science
Technical University of Lodz
Wolczanska 215
90-924 Lodz
Poland
E-mail: dzakrz@ics.p.lodz.pl

Liliana Byczkowska-Lipinska
Institute of Computer Science
Technical University of Lodz
Wolczanska 215
90-924 Lodz
Poland
E-mail: lilip@ics.p.lodz.pl

Ernestina Menasalvas
Facultad de Informatica
Universidad Politecnica de Madrid
Campus de Montegancedo s/n
28660 Boadilla del Monte Madrid
Spain
E-mail: emenasalvas@fi.upm.es

ISBN 978-3-642-42496-0 ISBN 978-3-642-02196-1 (eBook)

DOI 10.1007/978-3-642-02196-1

Studies in Computational Intelligence ISSN 1860-949X

Typeset & Cover Design: Scientific Publishing Services Pvt. Ltd., Chennai, India.

Printed in acid-free paper

9 8 7 6 5 4 3 2 1

springer.com

Preface

The overwhelming pace of evolution in technology has made it possible to develop intelligent systems which help users in their dayly life activities. Accordingly, methods of recording, managing and analysing data have evolved from the very simple file systems into complex ambient supportive intelligent systems.

This book arises as a compilation of methods, techniques and tools connected with data related issues: from modelling to analysis. A broad range of approaches such as database self-* techniques for ubiquitous environments, multimedia data, or data driven models will be reviewed. Different areas of applications, in which data models conceptualize nowadays reality, starting from e-learning to electric transformers will be considered.

The book is a collection of representative contributions to cover the spectrum related to data bases, which support decision making and data mining methods as well as conceptualization. Datawarehouse technology and modeling are presented in the first chapter together with the deep review of datawarehouse techniques for supporting e-learning processes with special emphasis on data cubes, all the tools are considered in the context of implementation of software application. The second chapter continues with the similar technology and deals with the community data warehouse architecture. Authors propose integrating a parallel query optimizer and grid architecture for query scheduling and optimization. Once database support has been reviewed, in the third chapter authors integrate datawarehouse techniques and data mining methods, for web query categorization, they concentrate on analysing the process and the database support required for its automatization. Subsequently, the deep survey of Fuzzy and Rough Sets methods for data analysis is presented in the chapter entitled Applications of Fuzzy and Rough Set Theory in Data Mining. Continuing with data analysis methods, in the fifth chapter, the author deals with the important topic of user driven modeling techniques. Data mining techniques, in particular clustering and sequential pattern mining are used for analysing navigational behaviour and learning styles of users. The following chapter refers to multimedia databases with focus on the techniques for image retrieval. After presenting supporting standards and solutions such as SQL/MM, Oracle and MPEG-7, the authors discuss using frequency domain techniques for content based image retrieval in multimedia databases. The last part of the book is more related to technology underlying database support. In the chapter entitled Query Relaxation in Cooperative Query Processing the authors deal with query mechanisms that enable to formulate meaningful queries by relaxing query constraints, they use semantic, structural and topological query relaxation. Special focus is done on similarities aspects and methodologies of obtained results evaluations. On the other hand, techniques for self-adaptation and interoperabolity are presented in the chapter entitled Ensuring Mobile Databases Interoperability in Ad Hoc Configurable Environments: A Plug-and-play Approach. Authors propose fully distributed agent-based architecture for building mobile database communities, taking into account physical mobility of hosts and logical mobility of database queries. The books ends by the presentation of the industry application, where data base is built according to requirements of expert system for monitoring and diagnosing in large power transformers.

We expect, that the book will be of interest for students, researchers and practicionners in any field related to development of intelligent systems, in which data bases and data analysis play crucial role. The editors would like to thank the contributors of the book for their effort in preparing the chapters. We would also like to give special thanks to Editor of the Serie prof. Janusz Kacprzyk for his help and his support and to prof. Mykhaylo Yatsymirskyy, for the book inspiration. Thanks to all the people who help in completing the book. We hope that our work will contribute to the evolution of databases research in the development of ubiquituos intelligent systems

March, 2009
Lodz, Danuta Zakrzewska
Madrid, Ernestina Menasalvas
Lodz, Liliana Byczkowska-Lipińska

Contents

List of Contributors

Ricardo Antunes
University of Coimbra
rantunes@dei.uc.pt

Angelo Brayner
University of Fortaleza (UNIFOR)
brayner@unifor.br

Liliana Byczkowska - Lipińska
Technical University of Lodz
lilip@ics.p.lodz.pl

Rogério Luís de Carvalho Costa
University of Coimbra
rogcosta@dei.uc.pt

Jitender S. Deogun
University of Nebraska-Lincoln
deogun@cse.unl.edu

Santiago Eibe Garcia
Universidad Politecnica de Madrid
seibe@fi.upm.es

Fernando Ferri
National Research Council Italy
fernando.ferri@irpps.cnr.it

Sergio Fialho
Federal University of Rio Grande do Norte
fialho@pop-rn.rnp.br

José de Aguiar Moraes Filho
University of Fortaleza (UNIFOR)
jaguiar@unifor.br

Pedro Furtado
University of Coimbra
pnf@dei.uc.pt

Patrizia Grifoni
National Research Council Italy
patrizia.grifoni@irpps.cnr.it

Maristela Holanda
Federal University of Rio Grande do Norte
mholanda@dca.ufrn.br

Dan Li
Northern Arizona University
Dan.Li@nau.edu

Ernestina Menasalvas Ruiz
Universidad Politecnica de Madrid
emensalvas@fi.upm.es

Bartłomiej Stasiak
Technical University of Lodz
basta@ics.p.lodz.pl

Arianna D'Ulizia
National Research Council Italy
arianna.dulizia@irpps.cnr.it

Eriko Werbet
University of Fortaleza
(UNIFOR)
eriko@unifor.br

Agnieszka Wosiak
Technical University of Lodz
agnieszka@ics.p.lodz.pl

Mykhaylo Yatsymirskyy
Technical University of Lodz
jacym@ics.p.lodz.pl

Danuta Zakrzewska
Technical University of Lodz
dzakrz@ics.p.lodz.pl

Marta E. Zorrilla
University of Cantabria
marta.zorrila@unican.es

Data Warehouse Technology for E-Learning

Marta E. Zorrilla

Abstract. E-Learning platforms are gaining popularity and relevance among organizations such as global enterprises, open and distance universities and research institutes. But regrettably these platforms present yet unsolved problems. One of these is that instructors cannot guarantee the success of the learning process because they lack tools with which monitor, assess and measure the performance of students in their virtual courses. Therefore, it is necessary to develop specific tools that help professors to do their work suitably. In this chapter, we show that data warehouse and OLAP technologies are the most suitable ones to build this software application. Likewise we explain the steps for its implementation from its conception up to the user interface development. Lastly, we summarize our experience in the design and implementation of MATEP, Monitoring and Analysis Tool for E-learning Platforms, which is a tool built in the University of Cantabria.

1 Business Intelligence Overview

In a market as competitive and global as the current one, information has become one of the main managerial assets. Until recently, the automated management of business processes (Invoice Management, Supply Chain Management, Enterprise Resource Planning, and so on) of companies and organizations by means of transactional systems had been sufficient to meet their information needs and do their work, but now they need to integrate and handle all this information in order to analyze it in line with the business aims that are to be achieved at every stage.

To help organizations with these tasks, at the end of the nineties, Business Intelligence (BI) tools appeared in the software market in order to enable the handling, consolidating and analyzing of large volumes of data, transforming these into valuable information for decision-making. In other words, these tools help analysts,

---

Marta E. Zorrilla
Department of Mathematics, Statistics and Computation,
University of Cantabria. Avenida de los Castros s/n 39005 Santander, Spain
e-mail: marta.zorrilla@unican.es

D. Zakrzewska et al. (Eds.): Meth. and Support. Tech. for Data Analys., SCI 225, pp. 1–20.
springerlink.com

managers and executives to have a more comprehensive knowledge of the factors affecting their business (metrics, key performance indicators, behaviour patterns, analysis of trends, etc.). Thus, they can modify their business processes in order to achieve targets such as increasing sales, reducing fraud, improving client relationship, increasing quality of the services offered and reducing costs.

In short, the Business Intelligence field provides companies with a framework to:

- Define and measure business key performance indicators (KPI) and understand their behaviour.
- Process, summarize, report and distribute the relevant information on time.
- Manage and share business knowledge with the organization.
- Analyse and optimise the processes that act on business key performance indicators.

In order to carry out these tasks, Business Intelligence solutions encompass a wide range of techniques and technologies: the data warehouse database as integrated repository of strategic information, the OLAP (On-Line Analytical Processing) technology for the exploration of information under different perspectives, dashboard, scorecard and reporting tools for the analysis and visualization of information and trends, and data mining techniques to discover meaningful patterns and rules in large volumes of data by automatic or semi-automatic means. Before continuing, it should be said that some authors, such as Kimball [22] consider that the "data warehouse" is the platform for business intelligence (DW/BI); however other authors such as Inmon [11] consider that the data warehouse is simply the database where the business data are consolidated and stored. In this work, we follow Kimball's idea.

Although Business Intelligence tools appeared to help companies tackle the most difficult business decisions, nowadays they are used in many other areas, such us in scientific applications (astronomy, bioinformatics, drug discovery, etc.), in governmental applications (surveillance, crime detection, profiling tax cheaters, etc.) or in the Web (search engines, e- commerce, web and text mining, e-learning, etc.).

An application inside this last area is explained in this chapter. This is a real BI solution, developed in the University of Cantabria, to monitoring and analysing the learners' behaviour in e-learning platforms, such as WebCT (now BlackBoard) [3], Moodle [20] or Claroline [4].

This chapter is organized as follows: First, we expose the problem that we want to solve using data warehouse technology and justify why this technology has been chosen. Next, we describe the components which make up the application. Then, we show the steps for its design and building, explaining each one of them in detail. Finally, we briefly summarize our experience and show the advantages that these technologies bring to our solution.

2 Why a Data Warehouse for E-Learning?

Nowadays, most universities, colleges and high schools are virtualising their teaching through e-learning platforms benefiting from their many advantages such as work "any-time, anywhere", use collaborative tools, support different styles of learning (collaborative learning, discussion-led learning, student-centred learning, resource-based learning), etc. However, these e-learning platforms do not cover all teaching aspects since they do not usually provide teachers with tools which allow them to thoroughly track and assess all the activities performed by all learners, nor to evaluate the structure of the course content and its effectiveness in the learning process ([18],[25], [33],[37]).

In fact, these environments provide the professor with access summary information such as the date of the first and the last connection, the number of visited pages according to the categories specified by the platform (not by the professor) or the number of read/sent mails by each student; the number of entries on each page and the average time spent on it per visit but, globally, per course and not by student. As can be deduced this information is not enough to analyze the behaviour of each student and his evolution. Even more, when the number of students and the diversity of interactions are high, the instructor has serious difficulties to extract useful information. Consequently, specific tools to help undertake this task must be developed. In our opinion, the use of business intelligence techniques applied to the Web ([10],[27],[32],[33]) helps us in getting our goal, given that these can generate statistics, analytic models and uncovered meaningful patterns from data.

In what follows, we mention some works that have been done in the e-learning field in order to solve these problems. CourseVis [18] is a tool that takes student tracking data collected by Content Management Systems and generates graphical representations that can be used by instructors to gain an understanding of what is happening in distance learning classes. It directly and exclusively uses web log files without building web sessions. GISMO [6] is another graphical interactive student monitoring and tracking system tool that extracts tracking data from Moodle. It provides different types of graphical representations and reports, such as graphs reporting the student's access to the course, graphs reporting all students' accesses to course resources, graphical representations of discussions pertaining to a course and graphs reporting data from the evaluation tools. Sinergo/ColAT [2] is a tool that offers interpretative views of the activity developed by students in a group learning collaborative environment. It integrates the information of user actions from log files with contextual information (events, actions and activities) in order to reconstruct the learning process. Mostow et al. in [21] describe a tool that shows a hierarchical representation of tutor-student interaction taken from log files.

On the other hand, using data mining techniques, we find particular solutions to specific goals. For example, Zaïane [34] suggests the use of web mining techniques to build an agent that could recommend on-line learning activities or shortcuts in a course web site based on learners' access history. Tang et al. [34] are working on a recommender system that finds relevant content on the web and personalizes and adapts this content based on the system's observation of the learners and the accumulated ratings given by them. TADA-ED [19] is a tool which tries to integrate various visualization and data mining facilities to help teachers in this discovering process. Lastly, an interesting survey about education data mining can be read in [25].

The solution that we propose tries to join and integrate both perspectives: to offer instructors the useful information by means of static and dynamic reports and graphically show them the discovered students' behaviour patterns. In order to store and manage all the data involved in the system, a solution supported in a data warehouse is considered as the most appropriate. There are several reasons which justify it.

First, the managed data (virtual courses, e-learning log file, demographic and academic data of students, admissions/registration info, and so on) are usually distributed in different data sources and, very often, the same information is gathered in a variety of formats, codes, etc. For example, the course code in the e-learning platform and the same one in the registration system. Even more, inconsistent data and duplicated registers can be found. For this reason, a new repository is required so that all the information is consistent, organized and standardized.

Second, the aim of each data source is not oriented towards making decisions but, on the one hand, to browse courses (e-learning platform) and, on the other hand, to register transactions (operational systems to manage admissions, registration, marks, etc.). This is why it is necessary to extract and transform data of interest for teachers from the different sources in order to calculate the key indicators with which they measure learners' performance and load them in a specific database to make decisions. It should be noted that the pages requested per each learner in the e-learning platform are usually registered in a log file supported as flat file, which requires to be previously processed in order to extract its information.

Besides, the fact of working with a database that follows the dimensional model will allow the system to quickly answer complex queries (accesses to several tables, aggregation and summarization operations, etc.). Likewise, the design and built of OLAP cubes is easier and simple.

On the other hand, a data warehouse is easily expandable and, thus, can also provide, in a future, with solutions to the needs of other actors of the educational environment. For example, site administrators could have parameters to improve the site efficiency and adapt it to the behaviour of its users; the academic management team could know the profile of their learners providing them with information to better organize their resources, both human and material, and their educational offer; and students could benefit from a personalized environment that will recommend them activities and/or resources that would favour and improve their learning.

Lastly, a data warehouse is, by definition, a specific information system to decision-making, driven by the needs of business users, fed from different data sources and built and presented from a simple perspective, which meets the requirements of this application.

In the following section the architecture of the BI solution for e-learning platforms is described.

3 E-Learning Data Warehouse Architecture

Figure 1 illustrates the architecture of our solution. As one can observe, the four components of a data warehousing environment [15] are present: data sources, data staging area, data area and data access tools.

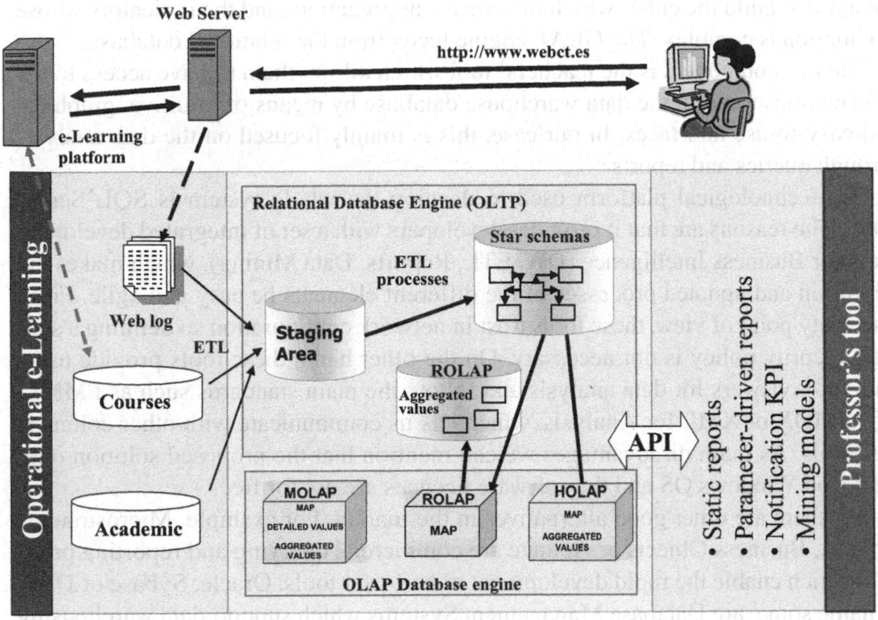
Web Server
http://www.webct.es
e-Learning
platform
Operational e-Learning
Relational Database Engine (OLTP)
Star schemas
ETL
processes
Web log
ETL
Staging
Area
ROLAP
Aggregated
values
Courses
Academic
API
MOLAP
MAP
DETAILED VALUES
AGGREGATED
VALUES
ROLAP
MAP
HOLAP
MAP
AGGREGATED
VALUES
OLAP Database engine
• Static reports
• Parameter-driven reports
• Notification KPI
• Mining models
Professor's tool

Fig. 1 E-learning Data Webhouse

The information of virtual courses, academic data as well as the e-learning platform log files which register the activity made in this tool (visited pages, date, time and name of student) are used as operational data sources. That means they are only queried in order to extract the information of interest.

Then come the ETL processes [14] which extract, transform and load this information onto the designed dimensional database (data area), and later also update it as a consequence of the necessary synchronization tasks between the operational and analytical systems. The ETL processes generally use a data stage area where data are transformed and consolidated. Besides, in this data stage area, the audit information (for example when the last load was done, how many rows were added or updated, ...) is written down and warehouse keys assignation is managed.

A Data area is where data are organized, stored, and made available for direct querying by users, reports and other analytical applications. The information is not only stored in detail but also summarized according to the different perspectives. For example, the admissions department of a university will need to know the number of students registered in each course in every academic year, whereas the professorship will be interested in knowing how this same number can be broken up by academic level or by age range.

These data can be stored either in a relational database management system or in a multidimensional system (OLAP), although generally both are used. In our proposal, the relational database is used to store the detail information and the OLAP database to build the cubes which include the aggregations and the indicators whose calculation is complex. The OLAP engine feeds from the relational database.

The last component is the teachers' tool which allows them to have access to the information stored in the data warehouse database by means of intuitive, graphical and easy-to-use interfaces. In our case, this is mainly focused on the data analysis through queries and reports.

The technological platform used to develop the whole system is SQL Server 2005. The reasons are that it provides developers with a set of integrated developing tools for Business Intelligence (DW, ETL, Reports, Data Mining), which makes the definition and updated processes of the different elements be easy and agile. From a security point of view, these tools trust in network configuration so defining a specific security policy is not necessary. On the other hand, these tools provide users with easy viewers for data analysis and follow the main standards such as PMML, SQL, MDX or XML for Analysis, which lets us communicate with other commercial tools. As main disadvantages we can mention that the proposed solution only works on Windows OS and the software licences are not for free.

But there are other good alternatives in the market. For example, Microstrategy, Cognos, Business Objects or Actuate are commercial querying and reporting products which enable the rapid development of end-user tools. Oracle, Sybase or DB2, to name some, are Database Management Systems which support data warehousing and OLAP technology. On the other hand, there is also an open source platform called Pentaho [24] which allows building the whole solution, but it requires a fair amount of customization.

4 E-Learning Data Warehouse Development Lifecycle

It is not easy to face the task of designing and developing a Business Intelligence solution for the first time. That is why it is advisable to have a process model which allows us to know the different tasks that have to be done and the deliverables that must be obtained. The proposed lifecycle (see Figure 2) is an iterative approach based on seven steps:

1. Identify business requirements and their associate value. This is generally gathered in a requirements document which will include, at the very least, a list of data elements, example questions and a list of desired reports that would help to answer the analytical questions. This can also include information about data sources (availability, constraints, quality) which will be used and, should it exist, a list of software and hardware requirements.
2. Design a single, integrated, easy-to-use, high performing information model that gathers the identified business requirements and build it physically in a relational database. It is recommended that this follow a dimensional schema.
3. Design the data stage schema and programme the ETL processes to load data in the dimensional schema. Likewise, implement the synchronization processes

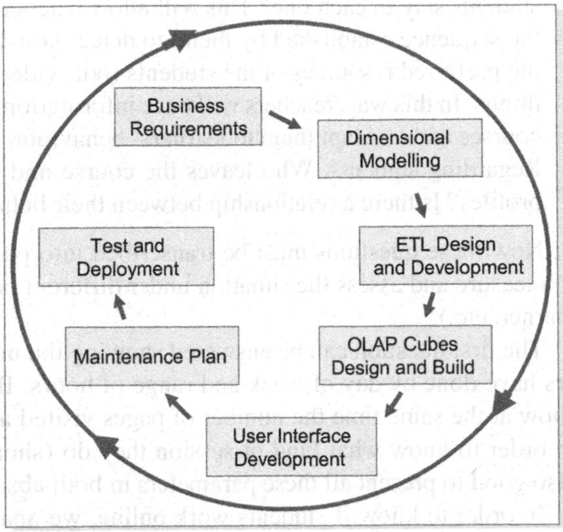
Business Requirements
Dimensional Modelling
ETL Design and Development
OLAP Cubes Design and Build
User Interface Development
Maintenance Plan
Test and Deployment

Fig. 2 DW/BI lifecycle

between data sources and the data warehouse database and, also develop the auditing system.

4. Design and build the OLAP cubes with the metrics, the complex calculations and the key performance indicators that the business experts require to make decisions.
5. Develop the user interface that allows business users to exploit the data (query tools, static and dynamic reports, scorecards, and so on) and extract the knowledge (patterns, rules).
6. Define the system maintenance plan and,
7. Test the solution developed and deploy it.

As in the development of any software solution, there are, of course, other transversal tasks to be carried out such as the definition and management of the project plan (phases, milestones, human resources, deliverables, risks and contingencies) and the selection of the development and exploitation tools.

Next, following the mentioned lifecycle, we are going to explain the design of our application step by step.

4.1 Business Requirements

As previously mentioned, the first step is to identify the business requirements that, in our case, can be summarized into answering questions such as:

- Regarding follow-up of the course: When do students connect to the system? Do they work online? Could the value of a session be measured in relation to learning objectives? This would help teachers to carry out continual assessment.
- Regarding the course: How often do they use collaborative tools? What are the sequences of visited pages in each session, in which order, and how long do

students stay in each one? This will allow teachers to discover if students follow the sequence established by them, to detect non-visited pages or to know which the preferred resources of the students (pdf, videotutorials, etc) are among other things. In this way, teachers will have information to modify the structure of their courses and to adapt them to learners' behaviour.

- Regarding students: Who leaves the course and when? What are the students' profiles? Is there a relationship between their behaviour and their qualifications?

Now these questions must be transcribed into parameters which allow teachers to measure and assess the situation under different perspectives (date, time, course, learner, etc.).

The first question can be answered showing the number of sessions which learners have done by day of week and range of hours. Besides, it would be suitable to show at the same time the number of pages visited and the minutes spent on them, in order to know what kind of session they do (short, long, ...). And it would be also good to present all these parameters in both absolute and average values.

In order to know if students work online, we analyse the time spent by session and the number and the kind of visited pages in a certain period of time. With this information, if the teacher asked for a work, proposed a connection to the forum or whatever other event, he will be able to assess the time that learners have dedicated to each activity.

To give a value to sessions in relation to the learning objectives that have to be achieved in them, it is necessary to classify every activity that a learner can perform in the virtual course into a category, according to its predominant characteristic (communication, evaluation, learning, additional information, etc.). Hereby, the assessment of a session is carried out "weighing" the activities performed according to the teacher's criteria who will have to assign a weight to every category depending on the characteristics of the specific course. In this way, if the course has n categories established, the value of the session will be obtained as:

$$ValueSession = \sum_{i=1}^{n} Weight_i \cdot ValueCategorySession_i$$

$$where, \quad \sum_{i=1}^{n} Weight_i = 1 \quad and \quad 0 \leq ValueCategorySession_i \leq 1$$

Regarding the use of collaborative tools, the number of sessions and the number of times by session where a collaborative tool has been requested will be enough to know how often they are used. The same will serve for any other category of pages (learning, evaluation, etc.).

The sequence of pages by session cannot be evaluated by means of a parameter but browsing the requested pages. This is why teachers must organise the pages of their courses according to work sessions, chapters or any other structure that allows them to compare the path that they designed with the one followed by their students. The user interface must facilitate this comparative.

Concerning students' drop-out, this can be detected showing the number of sessions and the average duration of these per learner and per group in a specific period of time in order to be compared with.

Lastly, once the course has finished, the teachers are usually interested in knowing if there is a relationship between marks and learners' behaviour. For this professor must analyse the following measures: number of sessions, average visit duration, average of sessions per week and marks. These same parameters but shown in relation to gender, age and degree of students allow professors to analyse the profile of their students.

Now, we know the parameters which professors want to observe and under what perspectives. This is summarized in Table 1. Next, we must design the dimensional model that allows us to gather this information.

Table 1 Instructor Key Performance Indicators

Parameters	Dimension	Level
N^o sessions (total and average)	Date	Year, Month, Semester, Week, Day of week, Day
N^o visited pages (total and average)	Time	Hour, Range of hours
N^o pages with error	Course	Name
Visit duration (minutes)	Page	Category, Subcategory, Name
	Learner	Name, gender, age, degree
	Session	CampusInside

4.2 Dimensional Modelling

The designed data warehouse database follows the dimensional modelling [13]. This has been selected because it is a broadly accepted technique. Its success is mainly due to the fact that it allows users to easily understand databases and software to navigate databases efficiently (query performance).

It is worth remembering that a dimensional schema is made up of a central fact table and its associate dimensions. It is also called star schema because it looks like a star with the fact table in the middle and the dimensions serving as the points on the star.

In our web house three stars have been designed: ActivitySession_Fact, ActivityPage_fact and Results_fact.

ActivitySession_Fact which gathers solely the clickstream data extracted from the e-learning platform logs. The level of detail or grain of this fact table is a row for each completed learner session. A learner session is defined as the time spent by a student since he or she connects to a certain course of the system until he leaves it.

This fact table contains the measures that teachers need to know: the number of visited pages, the number of pages with error and the time spent on the e-learning session. As can be observed, all these measurements or facts are numeric and additive, meaning they can be summed up across all dimensions. The fact table also contains the foreign keys to dimension tables, and its primary key which is usually autonumeric.

As dimension tables have been designed the following ones:

- Date_dim: it gathers each day of year with all its characteristics (number of day, month, day of week, week, year, and so on).
- Learners_dim: it collects the students' information, such as name, gender, degree, etc.
- Sessions_dim: it stores an identifier per each session, writing down the IP and if it belongs to an educational centre (campus inside).
- Courses_dim: it collects the virtual courses' information.
- Page_dim: it gathers each page of each virtual course and the estimate time of study and its classification, if the instructor indicates them.

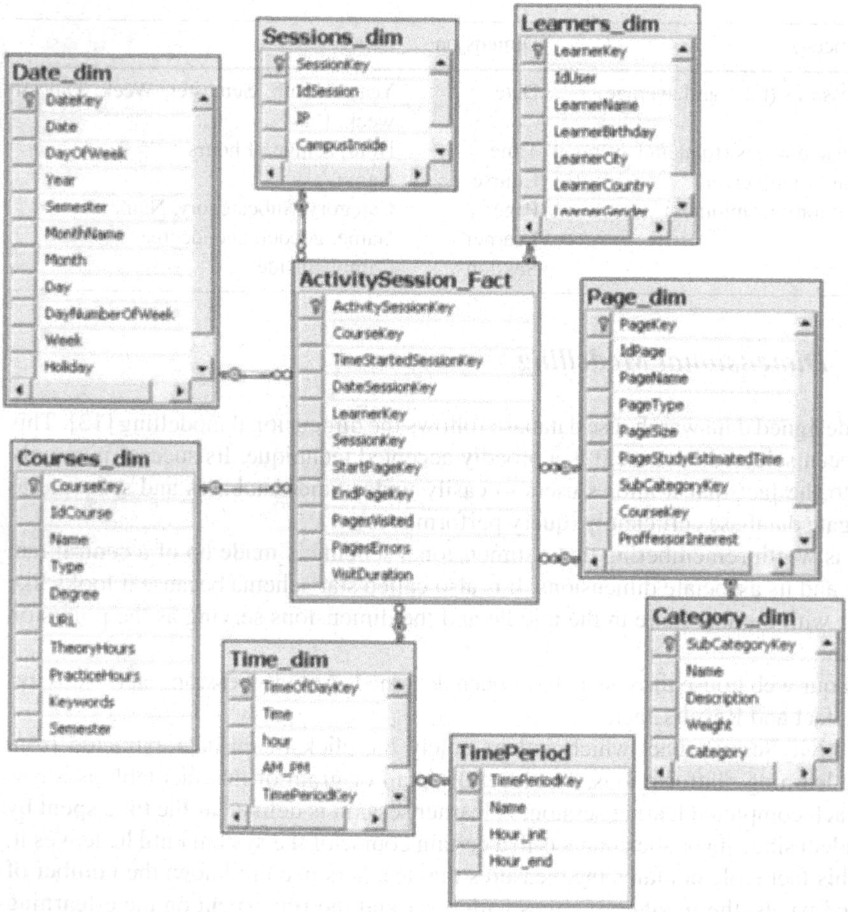
Sessions_dim
SessionKey
IdSession
IP
CampusInside
Learners_dim
LearnerKey
IdUser
LearnerName
LearnerBirthday
LearnerCity
LearnerCountry
Date_dim
DateKey
Date
DayOfWeek
Year
Semester
MonthName
Month
Day
DayNumberOfWeek
Week
Holiday
ActivitySession_Fact
ActivitySessionKey
CourseKey
TimeStartedSessionKey
DateSessionKey
LearnerKey
SessionKey
StartPageKey
EndPageKey
PagesVisited
PagesErrors
VisitDuration
Page_dim
PageKey
IdPage
PageName
PageType
PageSize
PageStudyEstimatedTime
SubCategoryKey
CourseKey
ProffessorInterest
Courses_dim
CourseKey
IdCourse
Name
Type
Degree
URL
TheoryHours
PracticeHours
Keywords
Semester
Category_dim
SubCategoryKey
Name
Description
Weight
Category
Time_dim
TimeOfDayKey
Time
hour
AM_PM
TimePeriodKey
TimePeriod
TimePeriodKey
Name
Hour_init
Hour_end

Fig. 3 ActivitySession_Fact schema

- Category_dim: it gathers the page classification defined by instructor of each course.
- Time_dim and Time_period: they store the time and its classification in periods of time.

It is worth mentioning that dimension tables are the foundation of the dimensional model, containing descriptive information relevant to analyse the fact table attributes from different perspectives. It can be observed that dimension tables have also got an autonumeric primary key and a column that gathers the natural key (for example, IdCourse in Course_Dim) of the operational system from which they come. This column is useful for the synchronization tasks.

ActivityPage_fact is another star (see Figure 4) with the same measures and dimensions but in this case, the grain is a row for each requested page per session. Furthermore, the fact table gathers the order in which these pages were requested.

Finally, Results_fact table (see Figure 5) summarizes the activity done by students in the course.

As can be observed, fact tables share the dimension tables. This requires that the dimensions be conformed.

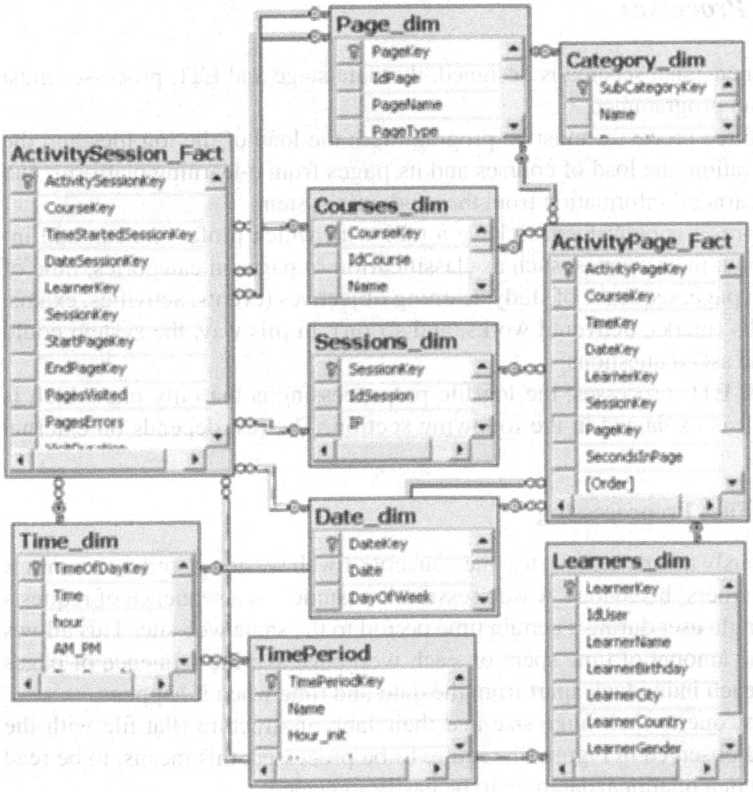
Page_dim
PageKey
IdPage
PageName
PageType
Category_dim
SubCategoryKey
Name
ActivitySession_Fact
ActivitySessionKey
CourseKey
TimeStartedSessionKey
DateSessionKey
LearnerKey
SessionKey
StartPageKey
EndPageKey
PagesVisited
PagesErrors
Courses_dim
CourseKey
IdCourse
Name
ActivityPage_Fact
ActivityPageKey
CourseKey
TimeKey
DateKey
LearnerKey
SessionKey
PageKey
SecondsInPage
[Order]
Sessions_dim
SessionKey
IdSession
IP
Date_dim
DateKey
Date
DayOfWeek
Time_dim
TimeOfDayKey
Time
hour
AM_PM
TimePeriod
TimePeriodKey
Name
Hour_init
Learners_dim
LearnerKey
IdUser
LearnerName
LearnerBirthday
LearnerCity
LearnerCountry
LearnerGender

Fig. 4 Fact constellation

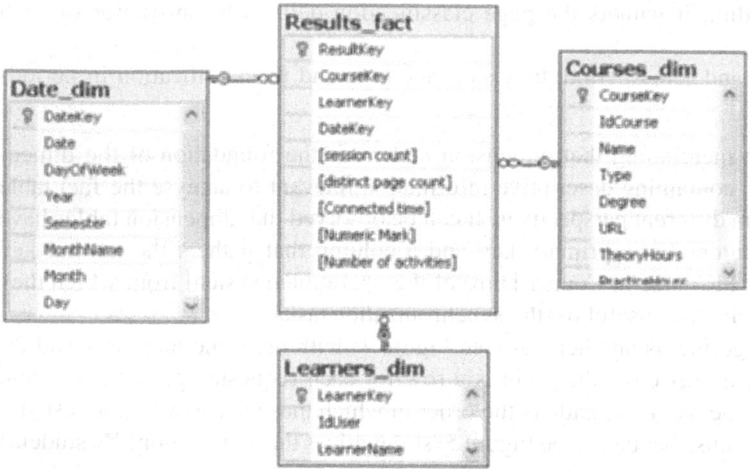
Results_fact
ResultKey
CourseKey
LearnerKey
DateKey
[session count]
[distinct page count]
[Connected time]
[Numeric Mark]
[Number of activities]
Date_dim
DateKey
Date
DayOfWeek
Year
Semester
MonthName
Month
Day
Courses_dim
CourseKey
IdCourse
Name
Type
Degree
URL
TheoryHours
Learners_dim
LearnerKey
IdUser
LearnerName

Fig. 5 Results_Fact schema

4.3 ETL Processes

Once the dimensional schema is designed, the data stage and ETL processes must be defined and programmed.

At least, three processes must be programmed: the load of the log files and the session generation; the load of courses and its pages from e-learning platform; and the load of learners' information from the academic system.

Furthermore, it is convenient to have a tool with which professors can add information about their courses, such as classification of pages in categories, time of study of each page, sequence of study, learning objectives (events, activities, exams, etc.) or results (marks, delivered works, and so on). In this way, the system could answer all the asked questions.

Among all ETL processes, the log file preprocessing is the only one which is generic and it is explained in the following section. The rest depends on external data sources.

4.3.1 Log Files Preprocessing

As we previously mentioned, we use the concept of web session in order to monitor and assess learners' behaviour. A web session is assumed as a sequence of requests made by a single user during a certain time period to the same web site. This allows us to have the amount of time spent on each web page and the sequence of pages accessed by each individual, apart from the date and time when it happens.

These files, due to their huge size and their lack of structure (flat file with the fields that are observed in Figure 6), require to be processed, this means, to be read and recorded in a relational database to be easily managed.

<ip_addr><ident><user><date><method><file><protocol><code><bytes><referrer><user_agent>

Fig. 6 Common Log Format specified as part of the HTTP protocol by CERN and NCSA [16]

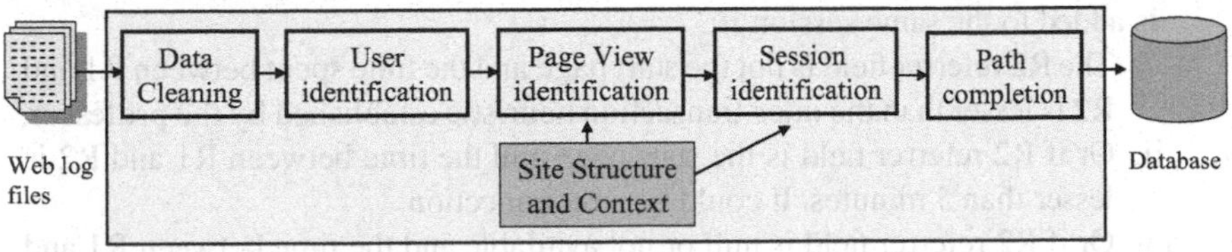
Data Cleaning
User identification
Page View identification
Session identification
Path completion
Site Structure and Context
Web log files
Database

Fig. 7 Steps in the pre-processing task

The pre-processing task is done according to [27] but, in this case, we adapt it to the educational environment. This includes the tasks of data cleaning, user identification, page identification, session identification and path completion (see Figure 7).

Data cleaning is the task of removing log entries that are not needed for the analysis process. In most cases, only the log entry of the HTML file request or a downloaded document is relevant and should be kept for the user session file.

User identification is the process of associating page references for each learner. Because the e-learning platform is an authenticate system, the name of the user is in the log file.

PageView identification takes the page and course which the user has viewed from the file field. That is why the site structure must be known.

Session identification takes all the page references for a given user and course in a log and breaks them up into user sessions. As the HTTP protocol is stateless, it is impossible to know when a user leaves the server, consequently, some assumptions have been made in order to identify sessions. In our particular case, the professor must indicate the time interval between two successive inter-transaction clicks according to his course organization and the time he estimates that their students will need to spend on each page. When this time is over, the session is considered finished.

Path completion fills in page references that are missing due to the browser and proxy server caching. In our case this step has not been developed because our architecture has not got a proxy server. With regard to local activities, such as backing and forwarding, these are not recorded in the log file.

In short, the process of sessions' generation can be summarized in these steps:

1. Filter rows with the following extensions in <file>: jpg, gif, bmp, png, css, asis, js,...
2. Filter rows with the IP of the server

3. Identify the course which each log row belongs to by means of extracting it from the <file> field
4. Order rows by IP, User, Course, [date and time] in ascendant order
5. Generate sessions:

    a. For each row, R1, with an IP-Course-User, a new session is generated.
    b. The following row, R2, must be read and if it has the same IP-Course-User, it is added to the same session if:
        i. The R2 referrer field is not the start page and the time spent between R1 and R2 is lesser than the inter-transaction heuristic established by the professor.
        ii. Or if R2 referrer field is the start page and the time between R1 and R2 is lesser than 5 minutes. It could be a reconnection.
        iii. Or if R2 referrer field is null or not available and the time between R1 and R2 is lesser than the inter-transaction heuristic established by the professor.

6. Estimate the time in the last page:

    a. For sessions with more than 10 pages, calculate the average time spent on pages of the same category in the session.
    b. For sessions with less than 10 pages, calculate the average time spent on pages of the same session.
    c. For sessions where the category page is not used, calculate the average time spent on this page in previous sessions.

4.4 OLAP Cubes

OLAP technology ([8],[12],[26],[30]) appeared in the 90s in order to meet the needs that managers and business analysts had to search accurate, update and complete information in their corporative systems (generally with large volumes of data) quickly, and handle and explore this information under different perspectives by means of simple and user-friendly formats.

Although relational databases are a great place to store and manage data, their internal data structures and the SQL query language have not been designed to answer complex business queries quickly.

To solve the performance problem, OLAP technology builds multidimensional structures called *cubes* that contain pre-calculated summary data (*aggregations*). In such a way that querying existing aggregated data is close to instantaneous compared to doing cold queries (no cache) with non pre-calculated summaries in place. For example, a typical aggregation would be adding up learners' sessions figures from month level, quarter and year.

Cube definition depends on the indicators and metrics that the end-user needs to observe and on the dimensions or perspectives under which he needs to assess them. In our application, three cubes were built to answer instructors' questions, one for each star, but a cube for each course or for each academic year or whatever other criteria can be defined in order to build manageable size data structures.

Next, we describe the steps to build the ActivitySession cube.

First, choose the ActivitySession_fact fact table which contain the measures to be included in the cube (PagesVisited, PagesError and VisitDuration) and, next, select the aggregation function to be applied to them. In this case sum for all the metrics.

Second, decide on the dimensions that the cube will include and define the hierarchies. In this case, the dimensions are: date, time, courses, sessions and learners.

In the date dimension, the following hierarchies were defined:

Day → Month → Year
Month → Semester → Year
DayNumber → Week → Year.

In time dimension, TimePeriod → Hour.
In page dimension, NamePage → SubCategoryName → Category.
And in learner dimension, LearnerName → CourseName.

Next, the following *calculated measures* are added to the cube: PageRequested as sum of PageVisited and PageErrors, Count of distinct learners and Count of sessions. These are needed to calculate the average time per session, the average number of pages per session and the average number of sessions per learner, which are defined as *calculated members.*

Calculated measures are stored in the cube, whereas only the definition of calculated member is cached in it. Their values are obtained in browsing time.

Once the dimensions and the cube are defined, they must be processed. This means, creating the multidimensional structure with all the aggregations defined. There are three alternatives: MOLAP, ROLAP and HOLAP.

MOLAP provides better query performance whereas ROLAP has the ability to handle large cubes only limited by the relational backend. HOLAP is a solution which balances performance cost and storage cost. MOLAP is advised for the data that are frequently queried.

4.5 User Interface

Once the whole data structure (data stage and data area) is built, one must continue with the development of the user's interface. There are different alternatives: one can purchase a packaged application and customize it so that data are taken from data area; or, use a commercial reporting and querying tool with which this interface can be built quickly and easily; or, otherwise, programme ad-hoc tools which allow end-users to interact with the strategic information.

As always, all alternatives have both advantages and disadvantages. The packaged applications present elaborate interfaces that meet the needs of business experts, but their customization is generally costly in terms of time and money. The development of ad-hoc tools allows the creation of flexible interfaces adapted to the users' needs but its average cost in time/person is higher and its implementation and deployment requires longer time. And the use of the commercial querying and

Course 05172

de 1 100% Buscar | Siguiente Seleccionar un formato

Year	Month	Week	Session Count	Avg. Pages/Session	Avg. Time/Session (min.)
⊞ 2005			12	17.92	2.36
⊟ 2006			2476	37.77	27.00
	⊞ January		10	68.80	6.27
	⊞ February		144	95.03	17.88
	⊞ March		657	48.71	35.09
	⊞ April		442	32.52	32.59
	⊞ May		709	33.06	30.00
	⊞ June		435	17.33	9.73
	⊞ July		21	18.19	2.51
	⊞ August		14	21.07	3.55
	⊞ September		44	25.09	26.22
	⊞ October			NaN	NaN

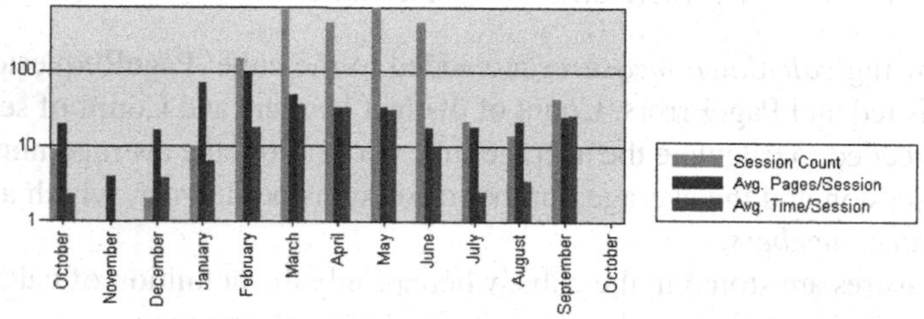
100
10
1
October
November
December
January
February
March
April
May
June
July
August
September
October
Session Count
Avg. Pages/Session
Avg. Time/Session

Course Introducción a los Métodos Multimedia: Desarrollo y Aplicaciones User uc15103

Year 2006 First Week February week 9

Last Week March week 12

de 1 100% Buscar | Siguiente Seleccionar un formato

MATEP USAGE ANALISYS OF COURSE: User vs Group UC UNIVERSIDAD DE CANTABRIA

Parameter	User	Group Average
Sessions	39	18
AVG Time/Session	37.918	33.86
AVG Pages/Session	72	52

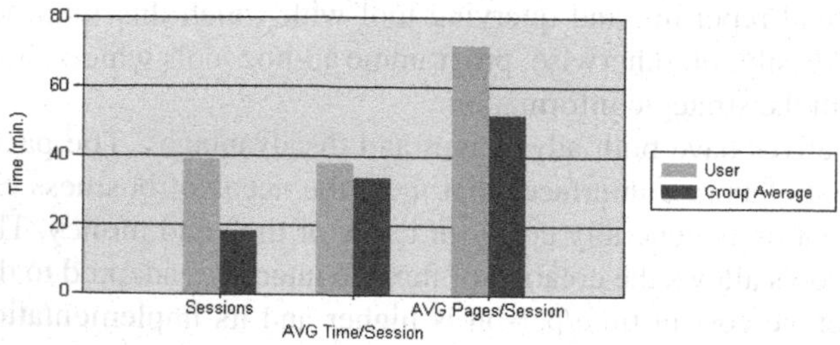
80
60
40
20
0
Time (min.)
Sessions
AVG Time/Session
AVG Pages/Session
User
Group Average

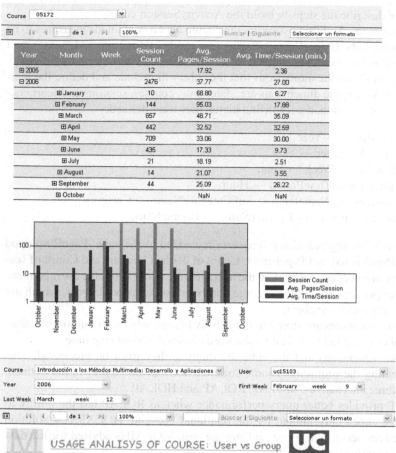

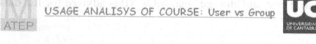

Fig. 8 Global usage course (top) and usage per learner (bottom)

reporting tools, which is the most commonly used option, provides programmers with many components that contribute to the rapid development of the applications although these tools present certain limitations to personalise the environment or add non-supported functions.

The spreadsheet programmes are another less sophisticated alternative, but also very used to manage exclusively OLAP cubes. They allow business analysts to explore and analyse data with a tool they are familiar with, although they lose the whole system perspective. They are suitable to work with the strategic data disconnected from the analytical system.

MATEP (Monitoring and Assess Tool for E-learning Platforms)([17],[38]) is the tool which we have developed for the instructor. This presents a Web interface and provides professors with a set of reports that give response to the teacher's requirements.

An example of these reports can be observed in Figure 8. It shows two reports which summarize the global usage analysis of the course (image at the top) and the usage per learner (image at the bottom). Both include the number of sessions, average time per session and number of pages per session measures and additionally, the second one also presents the average values of the course so that instructors can compare the measures. These reports are very useful because they allow teachers to evaluate the usage of their course, to detect if a student is close to giving up, if the students connect to the system frequently, or if the study effort is higher than expected.

As can be observed, the user establishes the conditions to do his analysis in the top area of the report. Furthermore, some reports allow him to analyze the indicators with different level of detail, making roll-up or drill-down in the report data area.

According to instructors' opinion, MATEP helps them to gain a more accurate knowledge of what is happening in their courses since it allows them to analyze and visualize data with different level of detail and perspectives, discovering student behaviour patterns' and understanding how their courses are used. That means that they have the quantitative and qualitative information available to take improvement actions about their courses.

5 Summary

This chapter describes an application of data warehousing and OLAP technology applied to the e-learning field. The application tries to solve the lack a face-to-face student-teacher relationship providing instructors with information with which they can track and assess their students' progress and evaluate the design and planning of their virtual courses.

After the development and implementation of the system, it can be said that two were the most demanding time tasks. The first one was the difficulty to recognize in the log file the page requested to the server due to how the e-learning platform writes down this information. Sometimes the same page appeared in the requested field with up to three different texts; depending on the way that learner had surfed to browse it. And the second one, to determine the heuristics with which to create web sessions, reaching the conclusion, after many tests, that because of the great variability existing in the composition of the courses (based on pdf files, videotutorials, html pages, etc) the teacher who designs and organizes the course is the one who can better determine them.

On the other hand, it has been proved that only the analysis of the clickstream allows instructors to have a good idea of what is happening in the course [35] but when its content is enriched with external information such as learners' academic and demographic data and information about the course, instructors can extract a more complete and accurate knowledge. An example of the conclusions that a teacher extracted from her virtual course using the tool MATEP can be read in [1].

Besides, it has to be said that the collaboration of the e-learning platform administrator and, in particular, of virtual courses' teachers has turned out to be essential for the development of the system. Teachers have had to do surveys, propose activities, ask learners directly, introduce the context information, take notes about the process and so on in order to have enough information to check and validate the goodness of the system.

Regarding technology, the fact that this application has been developed on a BI/DW architecture, has the following advantages:

- The application is independent from e-Learning platform to which it audits.
- The application gathers academic and demographic data as well as the context specified by instructor.
- The application provides instructors with information updated with the frequency that administrator establishes on the execution of ETL processes. Generally, on a daily basis.
- The system manages the information efficiently although the volume of data is high.
- The use of OLAP technology not only offers quick answers but it also allows users to manage the information: increasing the level of aggregation along one or more dimension hierarchies, decreasing the level of aggregation or re-orienting the multidimensional view of data.
- The system is easily extended to meet the needs of other users of the academic environment (students, management team, administrators, ...).
- Besides, this can be also used to store information that intelligent algorithms require to obtain patterns and to be the repository of these (for example users' behaviour and navigation patterns) [35].
- The modular architecture of the solution and the use of standards facilitate the change of any of its components.

Finally, it is essential that system/database administrators work closely with the Data Warehouse designers to better understand and characterize the database performance requirements.

References

1. Álvarez, E., Zorrilla, M.E.: Orientaciones en el diseño y evaluación de un curso virtual parala enseñanza de aplicaciones informáticas. IEEE-RITA 3(2), 1–10 (2008)
2. Avouris, N., Komis, V., Fiotakis, G., Margaritis, M., Voyiatzaki, G.: Logging of fingertip actions is not enough for analysis of learning activities. In: Proccedings of Workshop Usage Analysis in learning systems (AIED 2005), Amsterdam (2005)
3. Blackboard, `http://www.blackboard.com`
4. Claroline, `http://www.claroline.net/`
5. Cooley, R., Mobasher, B., Srivastava, J.: Data Preparation for Mining World Wide Web Browsing Patterns. Journal of Knowledge and Information Systems 1(1) (1999)
6. GISMO (2007), `http://gismo.sourceforge.net/`
7. Han, J.: Data mining: Concepts and Techniques. Morgan Kaufmann, San Francisco (2006)
8. Harinath, S., Quinn, S.: Professional SQL Server Analysis Services 2005 with MDX. Wiley Publishing Inc., Chichester (2006)
9. Hernández Orallo, J., Ramírez Quintana, M.J., Ferri Ramírez, C.: Introducción a la minería de datos. Pearson Prentice Hall, London (2004)
10. Hu, X., Cercone, N.: A data warehouse/online analytic processing framework for web usage mining and business intelligence reporting. International Journal of Intelligent Systems 19(7), 585–606 (2004)
11. Inmon, W.H.: Building the Data Warehouse. Willey & Son, Chichester (2002)
12. Jacobson, R.: Microsoft SQL Server 2000 Analysis Services step by step. OLAP Train. Microsoft Press (2000)
13. Kimball, R., et al.: The Data Warehouse Lifecycle Toolkit: Tools and Techniques for Designing, Developing, and Deploying Data Warehouses. John Wiley & Sons, Chichester (1998)
14. Kimball, R., Caserta, J.: The Data Warehouse ETL Toolkit. John Wiley & Sons, Chichester (2002)
15. Kimball, R., Ross, M.: The data warehouse toolkit: the complete guide to dimensional modelling. John Wiley & Sons, Chichester (2002)
16. Luotonen, A.: The common log file format (1995), `http://www.w3.org/pub/WWW/`
17. Martín Fraile, L.: Monitoring and analysis tool for e-Learning platforms. Final Degree Project directed by Zorrilla Pantaleón, M. University of Cantabria (2007)
18. Mazza, R., Dimitrova, V.: CourseVis: A graphical student monitoring tool for supporting instructors in web-based distance courses. International Journal of Human-Computer Studies 65(2), 125–139 (2007)
19. Merceron, A., Yacef, K.: Tada-ed for educational data mining. Interactive Multimedia Electronic Journal of Computer-Enhanced Learning 7(1), 267–287 (2005)
20. Moodle, `http://moodle.org/`
21. Mostow, J., Beck, J., Cen, H., Cuneo, A., Gouvea, E., Heiner, C.: An Educational Data Mining Tool to Browse Tutor-Student Interactions: Time Will Tell! In: Proc. of Workshop on educational data mining, pp. 15–22 (2005)
22. Mundy, J., Thorthwaite, W., Kimball, R.: The Microsoft Data Warehouse Toolkit: with SQL Server 2005 and the Microsoft Business Intelligence Toolset. Wiley Publishing Inc., Chichester (2006)
23. Pabarskaite, Z., Raudys, A.: A process of knowledge discovery from web log data: systematization and critical review. Journal Intelligent Information Systems 28, 79–114 (2007)

24. Pentaho, `http://www.pentaho.com/`
25. Romero, C., Ventura, S.: Data mining in E-Learning. Advances in Management Information, vol. 4. WIT Press (2006)
26. Spofford, G., et al.: MDX Solutions: With Microsoft SQL Server Analysis Services 2005 and Hyperion Essbase. Wiley Publishing, Chichester (2006)
27. Srivastava, J., Cooley, R., Deshpande, M., Tan, P.: Web usage mining: discovery and applications of usage patterns from Web data. SIGKDD Explor. 1(2), 12–23 (2000)
28. Tan, P., Steinbach, M., Kumar, V.: Introduction to data mining. Pearson Prentice Hall, London (2006)
29. Tang, C., McCalla, G.: Smart recommendation for an evolving e-learning system. International Journal on E-Learning 4(1), 105–129 (2005)
30. Thomsen, E.: OLAP Solutions: Building Multidimensional Information Systems, 2nd edn. John Wiley & Sons, Chichester (2002)
31. Witten, I., Frank, E.: Data mining. Practical machine learning tools and techniques. Morgan Kaufmann, San Francisco (2005)
32. Zaïane, O., Xin, M., Han, J.: Discovering Web Access Patterns and Trends by Applying OLAP and Data Mining Technology on Web Logs. In: Proc. Advances in Digital Libraries, Santa Barbara (1998)
33. Zaïane, O.: Web Usage Mining for a Better Web-Based Learning Environment. In: Proc. of Conference on Advantage Technology for Education, Alberta, Canada (2001)
34. Zaïane, O.: Building a Recommender Agent for e-Learning Systems. In: Proceedings of the International Conference on Computers in Education (ICCE) (2000)
35. Zorrilla, M.E., Menasalvas, E., Marín, D., Mora, E., Segovia, J.: Web usage mining project for improving web-based learning sites. In: Moreno Díaz, R., Pichler, F., Quesada Arencibia, A. (eds.) EUROCAST 2005. LNCS, vol. 3643, pp. 205–210. Springer, Heidelberg (2005)
36. Zorrilla, M., Millán, S., Menasalvas, E.: Data webhouse to support web intelligence in e-learning environments. In: Proc. of the IEEE International Conference on Granular Computing, Beijing, China (2005)
37. Zorrilla, M.E., Marín, D., Álvarez, E.: Towards virtual course evaluation using Web Intelligence. In: Moreno Díaz, R., Pichler, F., Quesada Arencibia, A. (eds.) EUROCAST 2007. LNCS, vol. 4739, pp. 392–399. Springer, Heidelberg (2007)
38. Zorrilla, M.E., Álvarez, E.: MATEP: Monitoring and Analysis Tool for e-Learning Platforms. In: Proc. of the 8th IEEE International Conference on Advanced Learning Technologies, Santander, Spain, July, pp. 611–613 (2008)

Optimizer and Scheduling for the Community Data Warehouse Architecture

Rogério Luís de Carvalho Costa, Ricardo Antunes, and Pedro Furtado

Abstract. In today's internet-connected data driven world, the demand on high performance data management systems is progressively growing. The data warehouse (DW) concept has evolved from a centralized local repository into a broader concept that encompasses a community service with unique storage and processing capabilities. This increase in popularity has lead to the appearance of new DW architectures and optimizations. In this chapter we propose two key inter-related enabler technologies for this vision: a parallel query optimizer which is able to optimize queries in any parallel DW independently of the underlying database management system (DBMS), and a scheduling approach for Grid DWs, which decides in which Grid site a query should be executed. We experimentally prove that the approaches allow the community Data Warehouse to work efficiently.

1 Introduction

In today's competitive world, Business Intelligence (BI) is considered to be an important asset to any enterprise which wishes to gain or maintain its market share. Data warehouses (DW) are usually considered to be a core element of any BI

Rogerio Luis de Carvalho Costa
Departamento de Engenharia Informatica - University of Coimbra,
Polo II - Pinhal de Marrocos 3030 - 290 - Coimbra - Portugal
e-mail: rogcosta@dei.uc.pt

Ricardo Antunes
Departamento de Engenharia Informatica - University of Coimbra,
Polo II - Pinhal de Marrocos 3030 - 290 - Coimbra - Portugal
e-mail: rantunes@dei.uc.pt

Pedro Furtado
Departamento de Engenharia Informatica - University of Coimbra,
Polo II - Pinhal de Marrocos 3030 - 290 - Coimbra - Portugal
e-mail: pnf@dei.uc.pt

D. Zakrzewska et al. (Eds.): Meth. and Support. Tech. for Data Analys., SCI 225, pp. 21–55.
springerlink.com

system, given the fact that they are very large business-related information repositories. DWs are typically used by decision makers to generate queries, make reports and perform analysis for the sake of decision support. Due to the colossal size of a DW and its constant need for high performing query executions, various specialized techniques were developed specifically for it. Amongst the available techniques, we mention the use of materialized views [7], special index structures [47, 48] and parallel systems [27].

With the world increasingly interconnected through the internet, the data warehouse concept has evolved to encompass also repositories for business or scientific "communities" that may be resident in a single location or in a world-wide setting. The sheer size of many data warehouses and the "community" view have both pressed the need for efficient, hardware and software-independent adaptable data warehouse platforms.

In this Chapter, we start off by centring our discussion on LAN-based parallel DWs, more precisely on parallel query optimizers. We analyze various parallel query optimizer architectures/solutions, and show that our optimizer, the Global Action Planner (GAP), is unique in the manner by which it tackles optimization. The novelty of the GAP is in the fact that it is able to generate parallel execution plans independently of the underlying database engine infrastructure. Parallelization is accomplished through the use of ordinary standalone database engines, without the need to resort to proprietary based parallel operations.

We then turn our attention to the usage of DW systems by geographically distributed global communities and organizations. In this case data distribution is in a much higher degree than the previously mentioned parallel systems. Not to forget that both the user and the DW system are WAN-connected. Data sharing in such global (real or virtual) organizations is becoming increasingly necessary [25]. In such environments, Grid Manager Software is usually used as a basic infrastructure for transparent remote resource access. Thus we discuss data placement issues and present a scheduling policy for the Grid-based DWs that aims at achieving user-defined Service Level Objectives (SLOs).

Both the GAP and the Grid-DWPA scheduler , are fundamental components of the Data Warehouse Parallel Architecture (DWPA) system. DWPA provides a run anywhere philosophy, allowing community DWs to operate in a networked environment, regardless of the context or the underlying hardware/software platforms. The GAP assumes the same run anywhere ideal, making it able to optimize site query processing independently of the underlying database infrastructure. When multiple sites are present the Grid-DWPA Scheduler interfaces with each local GAP , so as to evaluate costs and produce an efficient grid schedule.

The chapter is organized as follows: the next section contains pertinent background information on query processing and scheduling in parallel and distributed environments. Then, in Section 3, we present the Grid-DWPA architecture. Section 4 presents the GAP optimizer. We discuss its concepts and present experimental evaluation results. In Section 5, we discuss the use of grid-based DWs, and introduce the Grid-DWPA Planned Scheduling strategy. Finally, we draw conclusions in Section 6.

2 Background

Our approaches are concerned with offering top efficiency in both a hardware and software independent Data Warehousing platform for community Grids. In this section we review relevant background knowledge on parallel architectures, query optimizers in parallel and distributed settings and query scheduling on Grids.

Due to their high demand on storage and performance, large DWs usually reside within some sort of parallel system. These systems may assume various architectures. In [17] the authors describe three basic models by which a parallel storage system can be designed.

- The *Shared Nothing* (SN) architecture is composed of multiple autonomous Processing Nodes (PN) each owning their own persistent storage devices and running separate copies of the Database Management System (DBMS). Communication between the PNs is done by message passing through the network. A PN can be composed of one or more processors and/or storage devices.
- The *Shared Disk* (SD) architecture is characterized by possessing multiple loosely coupled PNs, similar to SN. However, in this case, the architecture possesses a global disk subsystem that is accessible to the DBMS of any PN.
- The *Shared Everything* (SE) is a system where all existing processors share a global memory address space as well as peripheral devices. Only one DBMS is present, which can be executed in multiple processes to utilize all processors.

From a hardware standpoint, SN systems are more cost effective than others because they are composed of commodity computers and an ordinary network. Furthermore, a large number of PNs can be interconnected in this manner because, other than the network, no resources are shared. The main disadvantage of SN architectures is that data needs to be interchanged between the nodes, which may lead to load-balancing issues and slower response times.

The Data Warehouse Parallel Architecture (DWPA) [27] is an environment adaptable and cost-effective middleware for parallel DWs. Its services are aimed at SN and mixed SN-Grid environments mostly because these are very scalable and their cost acquisition is relatively low.

In SN environments, the DW is partitioned amongst the PNs so that each DBMS instance can directly access data from its local partition. Partitioning is justified by the fact that PNs do not have enough storage space to accommodate the entirety of the DW, and that by subdividing the relations into various PNs the system is able to parallelize query executions. The DWPA architecture partitions a schema by using a partitioning strategy called Workload Based Placement (WBP) [27]. In short, WBP hash-partitions large relations based on the schema and workload characteristics of the DW, whilst small relations are replicated throughout the PNs. A relation is considered to be small if it can fit comfortably in physical memory, and the operations involving that relation are not significantly slower than those of a partitioned alternative. WBP tries to equi-partition relations whenever possible, so as to maximize query throughput. To equi-partition relations is to divide two or more relations using the same attribute and hash function. In doing so each PN is able to locally access and join the equi-partitioned relations, which avoids any unnecessary data exchange among the PNs.

DWPA can be organized into a set of node groups (NG) [29]. Each NG is composed of a set of PNs. NGs are usually created for reasons related to availability, performance enhancement or geographic locality. The usage of NGs benefits parallel query execution in various manners as they may contain any subset of the data warehouse data. The creation of multiple NGs, each containing the entirety of the DW, can also allow the execution of simultaneous queries in local or networked contexts, without these interfering with each others response times. Another application of NGs is to foment the use of bushy trees: if the DW is subdivided into two or more NGs then it is possible to segment a query and execute portions of it in separate NGs.

When confronted with a geographically disperse set of PNs, the DWPA middleware can organize these into sets of NGs. Considering a Wide-Area-Network (WAN) environment, these NGs can represent Grid sites. The Grid-DWPA is the system that results from applying DWPA in such settings. This specific configuration is confronted with issues concerning feasible data layout given present bandwidth and latency restrictions, site availability as well as efficient scheduling of queries into the sites.

There exist alternative software applications, to the DWPA middleware, that try to deal with parallel DWs in a different manner. The Data Warehouse Stripping (DWS) research project [14] applies a philosophy similar to DWPA, i.e. it does not need specific hardware to function, because it is able deliver query results using plain commercial off-the-shelf hardware. The DWS data placement strategy is fairly simple, partition the fact table using a round robin strategy and replicate the remaining tables throughout the PNs. Although good response times are achieved using single-fact pure star schemas, the architecture performs poorly with any other schema type/variation. The reason for this is because of its restrictive partitioning strategy, which does not foresee the partitioning of any other tables besides the main fact table.

One of the most popular commercial SN parallel database solutions is the IBM DB2 Parallel Edition [8]. Its partitioning strategy is similar to DWPA, in the sense that it hashes the partitioned relations so as to form collocated data whenever possible. The internal organization of the DW is also made through NGs. The application is very flexible, it allows database administrators to introduce or remove PNs into the system, without these having to spend a lot of time reconfiguring the DW. However despite the ease of maintenance, DB2 reveals some problems in adapting to heterogeneous and non-dedicated environments. A major setback of the DB2, as with any proprietary application, is the fact that complex solutions such as this require the acquisition of the entire software bundle, which is usually very costly.

Another popular commercial solution is the Oracle Real Application Cluster (RAC) [49]. The RAC assumes a hybrid parallel architecture. The PNs get their data from a high end disk array such as a Network Attached Storage (NAS) or a Storage Area Network (SAN), meaning that the application is partially built on SD. Once the data is loaded onto the PNs the cache fusion technology is activated, which basically allows the PNs to query each others cache so as to obtain needed data without resorting to the disk array. Oracle classifies the cache fusion as a form of SE. The main disadvantages of the RAC are its necessity for high end hardware such as a disk array, as well as a fast network interconnect for inter-node and disk array communication. Thus a fair amount of expenditure is needed to acquire both the hardware and software to build this system. Neither of the referred technologies are prepared to service requests within a Grid environment, as DWPA does.

2.1 *Query Optimization*

As database systems evolve, the need to develop query optimizers that are able to interpret their new features also arises. The evolution process has not been simple, seeing as query optimizers have to conjugate database feature compatibility with reasonable search times.

The first globally accepted standalone query optimizer was presented by Selinger et al. in [53] for the System R database engine. The optimizer was based on Dynamic Programming (DP), and tried to obtain the best execution plan by successively joining relations using the cheapest access paths and join methods. After each join iteration, the costliest plans were pruned leaving only the most eligible for the next phase. The exceptions to the previous rule were costlier plans considered to be *interesting*, because they could benefit future join plans or database operations that had yet to be accounted for. The System R query optimizer had a major disadvantage, its space and time complexities are exponential, which limited the join cardinality of a submitted query.

In the early 1980's researchers began to tackle query optimization in distributed environments. At the time standalone query optimizers were not equipped to handle features inherent to distribution, such as query segment node attribution or communication costs. Extensive work was done to adapt distributed concepts to the standalone optimizer domain. An example of this attempt is the R* [44] which adds distribution features to the DP algorithm (discussed earlier). Kossman and Stocker [40], develop a series of algorithms that adapt standalone DP to distributed environments. They reduce the inherent complexity, of such an adaptation, by introducing greedy algorithms that lessen the overall search space of a distributed query. Heterogeneity is also a major preoccupation in distributed query optimization, because slower sites influence the global response time. In [18], the authors propose optimizations that take into account alternative join sites by using data replication. The paper also describes a simple method for optimizing inter-site joins.

Parallel architectures can exploit two types of parallel features, i.e. *inter-query* and *intra-query*. The first tries to simultaneously execute independent queries by assigning them to different sets of processors. Intra-query parallelism breaks down the operators that compose a query, conveying them to different processors for concurrent execution. We can further subdivide intra-query parallelism into inter-operator and intra-operator parallelism. The latter promotes the division of a specific operator into fragments, and subsequent concurrent execution of each fragment by a separate processor. Inter-operator parallelism allows two or more operators to be simultaneously executed by different sets of processors. The introduction of these features augments the search space of a parallel system. Thus it is important that parallel optimizers be able to reduce the search space without compromising plan generation.

One of the first parallel query optimizers to appear, the XPRS [36], created effective parallel plans using a two-phase technique. The first generated a traditional single processor execution plan, which was then scheduled for parallelization in the second phase. However the authors did not study the effects of processor communication, which limited its applicability on architectures that heavily rely on this resource, i.e. SN. The Join Ordering and Query Rewrite (JOQR) algorithm proposed in [33] also employed a two phase approach. However the first phase was different from that of the XPRS. JOQR rewrote the queries, submitted to it, by using a set of heuristics to diminish the search space, and took into account processor communication by using node colouring [34]. Node colouring applies a colour to represent PNs with relations partitioned by the same attribute. Joining relations with the same colour would account for less communication costs, and thus diminish the global response time. In [54], the authors propose a one-phase search algorithm that uses graph theory to solve the best order by which to join a set of relations. In their approach they apply a CHAIN algorithm to try and minimize the response times of the execution plans. CHAIN is based on DP, yet it can eliminate more plans if Kruskal's [42] or Primm's [50] spanning trees are used as a greedy heuristic. The authors focus on equi-join scenarios leaving out the rest of the join types, this limits the applicability of the algorithm.

With the ever-growing complexity of today's queries, researchers are focusing on methods that try to diminish the margin of error of current query optimizers. Special interest has arisen in the usage Robust Query Optimization [3, 4, 13] techniques. The methodology allows an optimizer to decide whether it wants to pursue a conservative or more aggressive plan to resolve a query. The conservative plan is likely to perform reasonably well in most situations, whilst the aggressive plan (traditional method) can do better, if the estimations are reliable, or much worse if these are rough approximations. Other researchers have studied Parametric Query Optimization [30, 38, 39], which is based on the finding of a small set of plans that are suitable for different situations. The method postpones some of its decisions until runtime, using a subset of plans to decide how best to proceed to the next execution phase. Parametric query optimization does particularly well in scenarios where queries are compiled once and executed repeatedly, with possible minor parameter changes. However the subject that has captivated most of the recent interest is that of Adaptive Query Processing. The term classifies algorithms that use execution feedback as a means of resolving execution and optimization problems related to missing statistics, unexpected correlations, unpredictable costs and dynamic data. The subject is so extensive and diverse that we leave this discussion out of our chapter. In [16], the authors conduct a reasonably complete survey on the issue.

Having reviewed query optimization, we now move on to a more specific environment, that of grid-enabled data warehouses and grid query scheduling.

2.2 *Query Scheduling in Grids*

The dissemination of high-speed networks and the popularity of the Internet, in conjunction with the availability of low-cost powerful computers and storage systems, have lead to the possibility of masking highly-distributed resources into a single transparent system. This transparent layer can be provided by the Grid [5].

One can define the Grid as being an infrastructure that coordinates resource sharing among multi-domain users, allowing the usage of these resources by different applications transparently [9, 25]. The Grid coordinates the access to a wide range of possibly heterogeneous resources, like computers, supercomputers, clusters, storages systems and networks, which may belong to different organizations [9, 25].

As the resources may be shared amongst different organizations, there should exist an array of restrictive rules that define the usage profiles of each grid user. These rules may form limited access based on time frames to resource capacity usage constraints [25]. In fact, in a multi-domain grid, domain controllers may have a considerable degree of autonomy to impose domain specific rules: the resources of a grid are not subject of a rigid centralized control [5].

In such multi-domain environments, there are implementation challenges related to several aspects, like interoperability and security issues. Grid Resource Manager (GRM) Systems have appeared in order to deal with those challenges. GRM Systems usually provide several grid related services like: authentication, remote job execution and control, data transfer on the Grid, data replica catalog and job scheduling [5].

Job scheduling in the Grid is somewhat different than job scheduling in the more "traditional" parallel environments (like the Shared Nothing (SN) architecture) and offers various challenges. These challenges range from efficient network latency management (disperse resources can generate different latencies), to the detection and monitorization of resource heterogeneity, or even the supervision of domain specific resource usage rules and constraints.

In order to deal with the abovementioned matters, grid schedulers consider aspects that are not usually taken into account by centralized domain specific job schedulers. In Grid scheduling, the use of Quality-of-Service (QoS) related parameters when executing a job imposes Service Level Objectives (SLO) to a wide range of resources, like network and storage systems [52]. To achieve the desired level of service in Grid scheduling, it is common practice to employ resource reservation: as resources are shared among local and remote users, Grid schedulers can try to reserve the necessary resources to execute a job before assigning it to a processing node [52]. The establishment of Service Level Agreements (SLA) between service providers and service consumers is also common in the Grid environment.

The unique characteristics of the Grid environment, lead its scheduling architectures to adopt algorithms that are uncommon to single domain schedulers. According to [41], there exist three basic grid scheduling architectures:

- Centralized - is a scheduling model very similar to those used in single domain schedulers: all submitted jobs are sent to a central scheduler, which assigns their execution to the existing processing nodes. Centralized schedulers know which jobs have been assigned to each node. Being knowledgeable of the global scenario the schedulers are able reduce load imbalances and increase throughput. On the other hand this model hinders site autonomy and increases the risk of the scheduler becoming a bottleneck.

- Hierarchical - is a distributed architecture that also possesses a centralized scheduler. However this scheduler does not directly assign jobs to processing nodes. It assigns jobs to processing sites (cluster of processing nodes). Each site is composed of a Local Scheduler. The Local Scheduler interprets the inbound jobs and attributes them to the processing nodes of its site. In this architecture, each Local Scheduler has some degree of autonomy to implement domain specific policies, provided that it achieves the required service level objectives.
- Decentralized - in this model there exists no Centralized Scheduler, just a handful of Local Schedulers (one for each processing site), to which the jobs are submitted to. Each Local Scheduler is responsible for implementing the local domain specific rules for its assigned site. The advantage of this model is that the single point-of-failure, i.e. Centralized Scheduler, is eliminated from the system. In contrast, Local Schedulers make their decisions based on partial knowledge on the state of the system. Thus a large number of message interchanges may be necessary in order to increase the awareness a Local Scheduler has on the current load and state of all the sites.

Next we review briefly previous work on Resource Managers and Application-level Scheduling, which are basic knowledge related to our approach. We also review the more specific issue of database query processing on the grid, which plays an essential role in our solutions.

Grid Resource Manager Systems and Application Level Schedulers

Since the mid 1990s, various efforts at developing resource management systems for the grid and networked job scheduling systems have appeared. Applications such as Condor [57], Globus Toolkit [21] and Legion [32] try to manage job execution within highly distributed resources.

Condor is considered to be a first-generation GRM System [20]. It is a single-domain resource management system that uses remote idle machines to execute computing-intensive jobs. Condor uses *Classified Advertisements* (ClassAds) to schedule jobs. Each submitted job has its execution requirements published in ClassAds by an agent. Participant resources also use ClassAds, yet they use these to publish their capacities and usage constraints. A matchmaker scans job advertisements and resources advertisements searching for compatible requirements and capacities. Finally, any final details are negotiated by the agent of the job and the matched resource. When agreed upon, job execution commences [58].

Legion is a GRM system that creates a virtual machine abstraction of the resources of the Grid [32]. It is a framework in which every participant (including users) is modeled as an object. For example, application class objects are used to instantiate a Grid running application, whilst collections objects store information on the participant nodes, including processor type and Operating System (OS) version.

When instantiating a class object, it is possible to specify the execution requirements of the applications, these include the scheduler object used to schedule the execution of the application. Legion supports the use of multiple concurrent schedulers, which can be user-written. Collections can be accessed by schedulers to verify the state of the available resources [46].

Initially Legion provided a single default scheduler that verified the constraints of a job (parameters in class objects) and the available resources. It then randomly assigned the job to an eligible resource. Round-robin and performance-based schedulers were recently added to the system [46].

The Globus Toolkit (GT) is one of the most popular multi-domain GRM systems. Some even argue that Globus is a de facto standard for grid infra-structure software [22]. It is a collection of tools that provide grid-related basic capabilities, including a Grid Security Infrastructure (GSI) [24], a Resource Allocation Manager (GRAM) [15], a Monitoring and Discovery System (MDS) [19] and a Replica Location Service (RLS) [12]. The Community Scheduler Framework (CSF) has been incorporated to the system since GT version 3. It is a meta-scheduling framework that facilitates the use of user-specific scheduling policies. But application level specific schedulers could always use Globus services in order to implement job scheduling on the Grid. Condor-G [26] and Nimrod-G [10] are some of those schedulers.

Condor-G is an extension of the Condor resource manager system, which has been adapted to work with the Globus Toolkit. Nimrod-G is an extension of the Nimrod scheduler [1]. Both Condor-G and Nimrod-G use the Globus MDS to discover available resources and the Globus GRAM to initiate and monitor job execution. The Condor-G scheduling policy is similar to the standard Condor policy: it uses ClassAds and matchmaking to schedule jobs to available resources. The Nimrod-G scheduler tries to accomplish two kinds of user-defined constraints job deadline and execution price (Nimrod is an economic-based scheduler). In [60], the authors present an algorithm to schedule job execution in grids considering deadlines and available budget as user-defined SLO parameters.

Although the presented schedulers are considered successful, they are mostly oriented towards generic jobs. Data-involving query execution over the Grid could benefit by using more specific scheduling strategies, like our proposed Grid-DWPA's Planned Scheduling approach. We review next work on database query scheduling in Grids.

Database Query Scheduling in Grids

As was done with the migration of Condor and Nimrod from single-domain distributed environments to Grid distributed environments, the Polar* [56] distributed query processor has been incorporated into a higher level data integration architecture. In doing so it is able to schedule distributed grid-based query executions. Polar* is used as the base query scheduler for the OGSA-DQP architecture (Distributed Query Processor based on the Open Grid Services Architecture [23]) [2].

The OGSA-DQP runs on the Globus Toolkit and uses the OGSA architecture to build web services. OGSA-DQP provides two services: the Grid Distributed Query

Service (GDQS) and the Grid Query Evaluation Service (GQES). GDQS encapsulates the Polar* scheduler. Polar* is able to promote intra-query parallelism by partitioning the execution plan into individual tasks, that are then assigned to different processing nodes. Grid Query Evaluation services assure that the tasks are executed by the processing nodes. As the execution capacity of a grid may change over time, load balancing techniques are deployed to minimize imbalances [31].

Another matter to take into consideration, in query scheduling, is the data placement constraints of each site. Although data replication may be used to improve availability and performance, replicating or migrating data during query execution may not be a wise choice [51].

The OLAP-Enabled Grid [43] executes OLAP queries over grid-based DWs. The algorithm takes into consideration that a DW may have its data partitioned throughout several sites (all having the same database schema), and that each site can store cached data. The algorithm divides query execution into two phases: first, the scheduler tries to answer queries, using local cached data; if it cannot be done, then queries are sent to remote database servers, which have the necessary data to execute the query.

The usage of resident data to answer OLAP queries before sending it to remote sites is also discussed in [61]. In this case the data space is divided into uniquely identifiable chunks. A cost metric is used to construct the distributed execution plan. The metric assigns independent tasks to every candidate execution chunk. The final chunks are chosen on a least cost basis. In such work, the Globus Toolkit is also used in [61] as an underling infrastructure to OLAP query execution over the grid.

The strategies proposed in [43, 61] reduce data movement over the Grid. However they increase the processing costs of all local sites, needlessly wasting the available processing power of some remote resources. In certain situations, they can lead to local node thrashing.

3 Grid-DWPA Architecture

The Grid-DWPA is an architecture for deploying large data warehouses over network connected environments. It can be used to implement both global grid-based as well as domain-specific parallel DWs. Its main components are (also represented in Figure 1):

- **The Grid-DWPA Community Scheduler:** When multiple sites are present, the Grid-DWPA Community Scheduler assumes the responsibility of scheduling query execution among the available sites, choosing which site should execute each query considering Service Level Objectives (as it will be discussed in Section 5). The community scheduler interacts with the various site instances of the GAP Optimizer, as represented in Figure 1.
- **The GAP Optimizer:** Each grid-site runs an instance of the GAP. The GAP is a parallel query optimizer which optimizes queries independently of the underlying database management software. To obtain engine independency, our

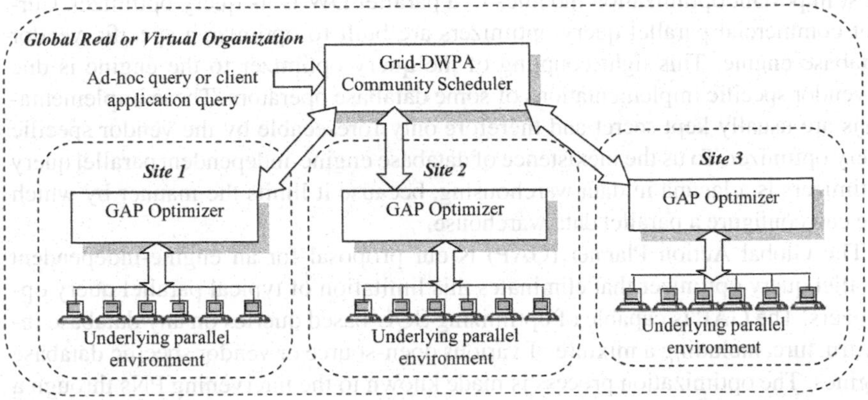
Global Real or Virtual Organization
Ad-hoc query or client application query
Grid-DWPA Community Scheduler
Site 1
GAP Optimizer
Underlying parallel environment
Site 2
GAP Optimizer
Underlying parallel environment
Site 3
GAP Optimizer
Underlying parallel environment

Fig. 1 Grid-DWPA Architecture Overview

optimizer constructs actions plans during query optimization and uses the SQL language to interact with existent DBMSs.

In multi-site implementations, the GAP*TE* (GAP *Time Estimator*) component is used to provide the query execution time estimations that are used by the Community Scheduler to choose which site should execute each query.

When used together, the Grid-DWPA Community Scheduler and the GAP Optimizer provide an efficient, and platform independent architecture to run a parallel DW anywhere. In Figure 1, these components are represented as part of a global data warehouse with three distinct sites. The Grid-DWPA Community Scheduler and the GAP Optimizer are organized in a hierarchical scheduling model. A single instance of the Community Scheduler interacts with each of the site GAP Optimizer instances, in order to schedule the query execution.

In the next section, we present the GAP Optimizer, which details the algorithm used to construct the engine independent action plans. The scheduling strategy used by the Community Scheduler is discussed later in this Chapter.

4 Generic Query Optimizer for DWPA

The Data Warehouse Parallel Architecture (DWPA) middleware was designed to service any shared nothing (SN)-based parallel Data Warehouse (DW) independently of its hardware and/or software infrastructure. This means that DWPA is capable of adapting to an array of DW configurations, ranging from executing over different proprietary database engines, to successfully detecting and using heterogeneous Processing Nodes (PN). This malleability is the product of a high level proprietary independent implementation, which allows our middleware to run anywhere.

DWPA is efficient because it applies a panoply of optimization algorithms, which constantly try to improve the overall performance of the parallel DW. One of the most important optimization modules of a parallel DW is its query optimizer. Current commercial parallel query optimizers are built to optimize a specific vendor database engine. This tight coupling of the query optimizer to the engine is due to vendor specific implementations of some database operators. These implementations are usually kept secret and therefore only foreseeable by the vendor specific query optimizer. To us the inexistence of database engine-independent parallel query optimizers is a lacuna in data warehousing, because it limits the manner by which one can configure a parallel data warehouse.

The Global Action Planner (GAP) is our proposal for an engine-independent parallel query optimizer that eliminates this limitation of typical parallel query optimizers. The GAP is capable of optimizing SQL-based queries on any database infrastructure, including a mixture of various open-source or vendor specific database engines. The optimization process is made known to the intervening PNs through a series of manipulated SQL queries. These queries indicate to the PNs how the GAP wants the user-given query to be processed.

The following subsections explain essential GAP related optimization concepts as well as the GAP algorithm.

4.1 GAP Concepts

Current parallel execution plans are composed of sets of database operations that must be executed in their presented order. Figure 2 illustrates an execution plan directed to a specific PN in the parallel DW. From the tree, one can see that all operations and their execution paths are predetermined.

The main setback of employing traditional parallel execution plans is that each proprietary parallel database solutions vendor uses its own protocol to communicate its optimizations to the PNs. This makes it impractical for an engine-independent query optimizer, e.g. GAP, to adopt such a structure. As an alternative we have created an optimization structure which is engine-independent, we call it the *action*

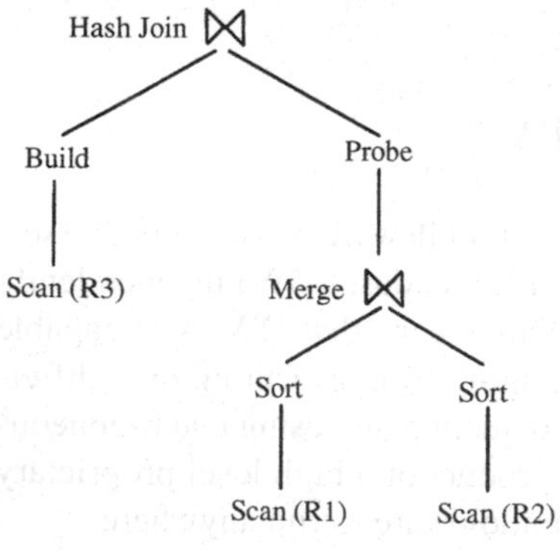
Hash Join
Build
Probe
Scan (R3)
Merge
Sort
Sort
Scan (R1)
Scan (R2)

Fig. 2 Excerpt of a traditional parallel execution plan

plan. An action plan is able to obtain engine independence by using the SQL language to communicate its optimization instructions to the PNs.

Local Query Actions

Local Query Actions (LQA) can be interpreted as being objects that contain information pertinent to the resolution of a fragment of a user-given query. The most important element of an LQA is its *local query*. A local query is an SQL-based statement. Its purpose is to indicate which relations a particular PN should execute in order to solve a fragment of the user-given query. It should be noted that local queries are only formed if the needed relations are physically present within a targeted PN. One could interpret our local query as being analogous to the site subqueries generated by ordinary distributed query optimizers, see [18]. However there are relevant differences between both. For example subqueries generated by a distributed query optimizer have not undergone a parallelization process, meaning that the subqueries are not parallelizable. In [28], the author enumerates a series of SQL transformation hints, which allow a user-given query to be broken down into a set of smaller efficient parallel queries. The most important hints focus on filtering and GROUP BY push downs, ORDER BY postponements, and most importantly aggregation function decompositions. The latter explains how some aggregation functions can not be directly calculated by parallel PNs, because this will result in miscalculation. To overcome the problem, aggregation functions are broken down into simpler parallelizable aggregation functions that will, in the end, provide the wanted answer.

Communication Actions

Communication actions (CA) are used to indicate a data exchange between a set of PNs. In an SN-based architecture each PN has a chunk of the DW allocated to it for local processability. In some situations a PN may need data from one or various other PNs in order to locally process its chunk. There exist two scenarios in which a PN will want to exchange data, it either needs an entire dataset from a specific PN, i.e. broadcast, or it requires various sets of tuples, from different PNs, i.e. repartitioning.

Figure 3 exemplifies a network broadcast from PN1 to the remaining PNs. After the broadcast, the exchanged dataset is used by the PNs to process a specific database operation. In Figure 4 relation F is hash-partitioned by attribute a whilst P is hash-partitioned by attribute b, the numbers beside the partition attributes correspond to the partition fragment. To be correctly joined both relations have to be partitioned by the same attribute and the same hash interval. In this example, F is repartitioned by attribute b, i.e. tuples in F are rehashed using the hash function applied to P. Rehashing can be interpreted as a migration of F's tuples from their previous hash-bucket to the new hash-bucket. In most cases this migration involves moving tuples to another PN.

Whenever the GAP encounters the need to exchange data between PNs it will use one of the abovementioned actions to do so.

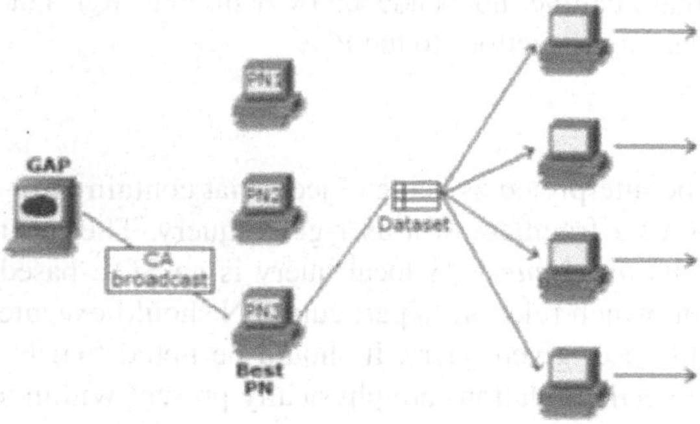
GAP
CA broadcast
Dataset
Best PN

Fig. 3 Dataset broadcasting

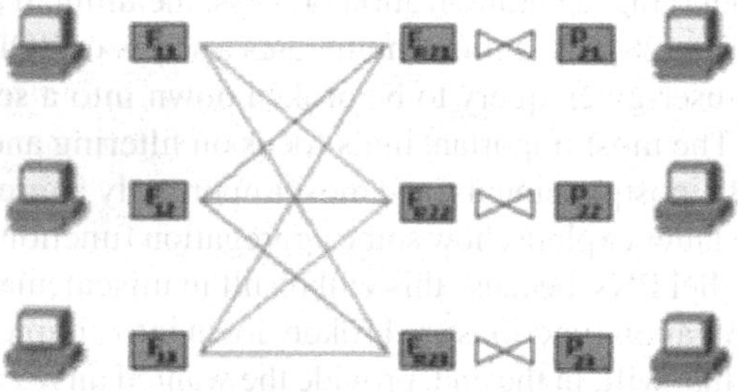

Fig. 4 Dataset Repartitioning

4.2 GAP Algorithm

To be able to optimize a user-given query the GAP must possess an efficient search algorithm. We chose to implement our parallel optimizer using a DP based algorithm because, of all the literature proposed optimization algorithms, DP algorithms generate the most efficient execution plans. However their efficiency normally results in medium to high memory consumptions and execution times. To overcome this handicap we have adopted some of the findings published by in [40]. The authors propose a set of variations to the classic DP, which allows us to maintain a high degree of confidence in the generated execution plans, while drastically reducing the memory and time consumptions.

Figure 5 illustrates the GAP algorithm in pseudo code. The algorithm starts off by receiving four input variables: the query that is to be optimized, i.e. q; the maximum block size k, which is used as a cut off heuristic (explained later); the capacity statistics which contain hardware performance values that determine the heterogeneity of the PNs; the list of the NGs that compose the DW.

Being based on DP, the GAP finds its action plans in a bottom-up fashion. All bottom-up search algorithms start by generating access paths, i.e. base points by which the remaining plans are constructed upon. In GAP we refer to these initial

Input: Query q on relations $D_1,...,D_n$, maximum block size k, capacity statistics for each PN (CE_{PNi}), nodegroup listing (NGL)
Output: A global action plan for q

```
1:   loadDataWarehouseConfig(NGL)

2:   for i = 1 to n do {
3:        generateBasePlans(D^i);
4:   }

5:   toDo = {D1, ..., Dn}

6:   while |toDo| > 1 do {
7:        k = min{k,|toDo|};
8:        for i = 2 to k do {
9:             for all S ⊆ toDo such that |S| = i do {
10:                 plan(S) = Ø;
                    mergePlans(S);
11:            }
12:       }

13:       find P, V with P Є plan(V), V ⊆ toDo, |V| = k such that
               eval(P) = min{eval(P') | P' Є plan(W), W ⊆ toDo, |W| = k};
14:       generate new symbol: T;
15:       plan({T}) = {P};
16:       toDo = toDo – V U {T};

17:       for all O ⊆ V do
18:            delete(plan(O));
19:  }

20:  finalizePlans(plan(toDo));
21:  prunePlans(plan(toDo));
22:  mergePlan(plan(toDo));
23:  return plan(toDo);

     where: S, O and V can be interpreted datasets
     P, T are plans with generated actions
```

Fig. 5 GAP algorithm

access points as base plans (lines 2 to 4). Each base plan contains within it a set of LQAs, i.e. one LQA for each intervening PN. The local query residing in the LQAs, of a specific base plan, references a single relation from the user-given query, as well as all SQL expressions related to the targeted relation, e.g. WHERE predicates, GROUP BY attributes, amongst others.

Figure 6 depicts a user-given query being decomposed into three distinct base plans (top left query). Queries shown in the top right section and bottom row of the figure are templates of the queries one would find within base plans R_1, R_2 and R_3 respectively. We refer to these local queries as templates because the name of the

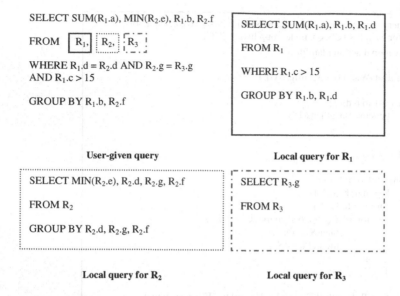

```
SELECT SUM(R1.a), MIN(R2.e), R1.b, R2.f
FROM R1, R2, R3
WHERE R1.d = R2.d AND R2.g = R3.g
AND R1.c > 15
GROUP BY R1.b, R2.f
```

User-given query

```
SELECT SUM(R1.a), R1.b, R1.d
FROM R1
WHERE R1.c > 15
GROUP BY R1.b, R1.d
```

Local query for R_1

```
SELECT MIN(R2.e), R2.d, R2.g, R2.f
FROM R2
GROUP BY R2.d, R2.g, R2.f
```

Local query for R_2

```
SELECT R3.g
FROM R3
```

Local query for R_3

Fig. 6 Example of a base plan decomposition of a user-given query

relation referenced in their FROM clause may vary from PN to PN. This variation only happens if the relation has been partitioned, e.g. relation $R1$ is partitioned into 3 PNs, the partition names could be R_{1a}, R_{1b}, and R_{1c}.

After all base plans have been created, the search algorithm begins its pairing strategy (lines 6 to 12). Pairing is an iterative cycle that successively combines base plans or sub-plans [1] with each other, until all relations referenced in the user-given query have been joined. In most cases plans are paired whenever the user-given query has a join clause that directly associates them, i.e. the relations referenced within the paired plans are associated through the join statement. Whenever relations mentioned in plans to be paired can not be directly joined, e.g. they are not equi-partitioned or not present in the joining PNs, then a CA is generated and added to the actions list of the resulting paired plan.

When necessary, the sub-plans are submitted for cost calculation. The GAP calculates the cost of a specific sub-plan to verify if it is worth maintaining in memory, if so then it will be used in the next pairing phase, if not it is discarded and no posterior sub-plans are based on it. The GAP estimates the cost of actions by using its Local Optimizer, i.e. GAP-LO, module to evaluate the local queries within the LQAs , and its network cost formulas to assess the CAs.

The GAP-LO is a standalone query optimizer that generates execution plans in much the same way a traditional proprietary query optimizer does, see Figure 1. However the GAP-LO does not know, and does not need to know, how a specific database engine implements its database operations, so it assumes a generic

[1] A sub-plan is the result of a pairing of two base plans, a base plan with a sub-plan, or two sub-plans.

implementation of the operations. The idea is to generate execution plans that are qualitatively similar to the execution plans generated by the proprietary optimizers. From its execution plan the GAP-LO will extrapolate a cost. This cost is then used by the GAP to differentiate low-cost plans, i.e. good plans, from high-cost plans.

The network cost formulas can also influence the "goodness" of a plan. In this case the operation, i.e. repartition or broadcast, is analyzed and a cost is given based on the number of intervening PNs and the amount of data to be transferred.

Knowing that the pruning of high cost plans may not be sufficient to significantly reduce memory consumptions and time execution, the GAP has imported the maximum block size value (k) algorithm, described in [40]. Parameter k indicates how many base plans or sub-plans, can be paired together before a pruning heuristic is triggered. When triggered, the heuristic compares all k-sized sub-plans and chooses the one with the least cost (line 13). All plans with the exception of the best k-sized plan, and the base plans not yet joined to the k-sized best plan are eliminated (lines 15 to 18). This process is iterative, and the greedy heuristic is activated whenever a multiple of k is reached (lines 7 and 8). By doing so the GAP is able to drastically reduce the amount of plans in memory, and seeing as there are less plans to combine, the GAP is able to reduce optimization time. k does not necessarily have to be directly input, e.g. generate an action plan with a maximum k of 3. It can be derived from a user indication stating how long he/she wants the optimization process to take, the shorter the optimization time the smaller the k. k can also be triggered whenever the optimizer is about overflow a pre-established memory threshold.

At the end of the optimization process the GAP will elect a best action plan, once again basing its choice on cost efficiency (lines 20 to 23). The chosen action plan will contain a sequence of actions which are passed to the DWPA middleware for processing. Figure 7 illustrates the execution process of a parallel action plan. As can be seen, the GAP does not send execution plans to the PNs, instead it forwards to them the local queries (SQL). These local queries are then optimized by the proprietary query optimizers of the database engines of the PNs. The GAP delegates this responsibility to the query optimizers, because they are more knowledgeable of the inner-workings of their database engines, and therefore have greater probabilities of generating better standalone execution plans than the GAP-LO. We call this strategy piggyback optimization. The optimization not only increases the certainty of faster response times by the individual PNs, but it also confers database engine independency to the GAP, seeing as local queries are used to convey the "execution plan".

4.3 Contributions to Parallel Optimization

As was mentioned in the background section, current research is focused most of all on the improvement of the plan generation process of evermore complex queries. Commercial products have adopted methods such as query rewriting [6, 35, 37], to obtain efficient query statements, mass parallelization methods, e.g. parallel pipelining [6, 8], and heterogeneity detection [6], for better resource handling. Academia and industry research has recently started to develop feedback mechanisms [16],

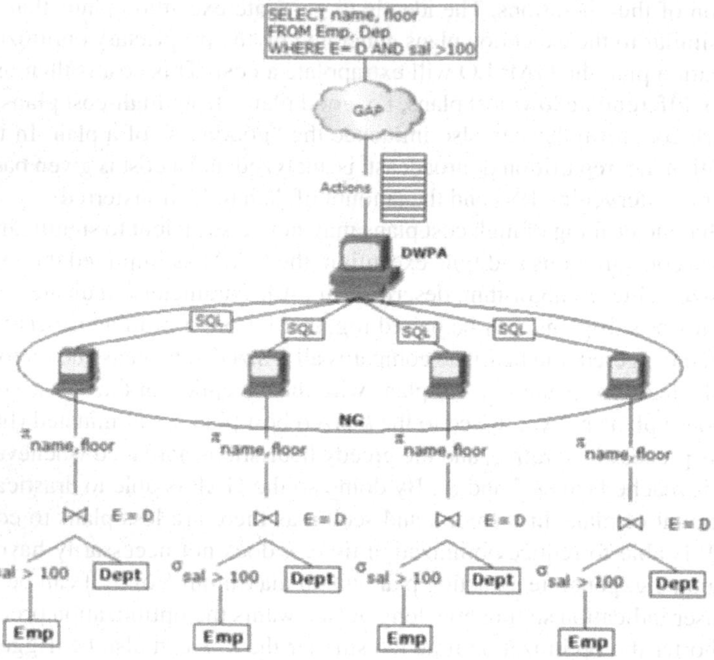
SELECT name, floor
FROM Emp, Dep
WHERE E= D AND sal >100
GAP
Actions
DWPA
SQL
SQL
SQL
SQL
NG
Dept
Emp

Fig. 7 Action Plan distribution

i.e. Adaptive Query Processing, that allow optimizers to dynamically correct their execution plans.

We consider the previously mentioned research important to the improvement of query optimizers in general. However we believe that there exists an area that has not received attention, in either industrial or academic circles. Current parallel query optimizers are tightly-coupled to the database engines they optimize. Most developers have done so because they believe that to be efficient a query optimizer must be knowledgeable of how database operations are implemented on a specific database engine. This limits the usage of an optimizer to just that database engine. We believe that by implementing a novel engine independent architecture such as GAP, it is possible for parallel optimizers to effectively generate execution plans for more than one type of database engine. This not only allows developers to free the optimizer from the database engine, but also allows data warehouse owners to compose their system using various core technologies from different sources.

4.4 GAP Results

In this section we show some of the achieved results when using the GAP in a parallel environment. We prove that the GAP is able to successfully promote intra-operator and inter-operator parallelism, using open-source database engines that possess no

Table 1 TPC-H Relation Cardinality for SF=10

Relation	Cardinality	Size(GB)
Lineitem(LI)	60.000.000	9.89
Orders(O)	15.000.000	1.96
Partsupp(PS)	8.000.000	1.25
Part(P)	2.000.000	0.44
Customer(C)	1.500.000	0.27
Supplier(S)	100.000	0.02
Nation(N)	25	0.00
Region(R)	5	0.00

parallelization features. Finally we demonstrate that the GAP is able to use alternative data placement layouts to improve the generation of its global action plans.

The generic configuration of the parallel data warehouse is indicated on an experiment to experiment basis. However the PNs mentioned in the experiments have the following hardware configurations:

- CPU - 122MIPS
- RAM - 512MB
- Disk size - 40GB, running at 7200RPM
- Network - 100Mbps
- DBMS - PostgreSQL8.2

All queries executed in the experiments are TPC-H based queries [59], some of these have been altered so as to better illustrate the objective of the experiment. The TPC-H schema is configured to a scale factor (SF) of ten. Table 1 shows the approximate size and cardinality that each relation occupies.

Schema partitioning is explained on an experiment to experiment basis.

Intra-operator Parallelization

The following test shows that the GAP is capable of generating parallel action plans using intra-operator parallelism. To prove this we setup two NGs. NG_1 has just one PN which holds the entire TPC-H schema. NG_2 has five PNs which have equally partitioned between them the largest relations of the TPC-H schema.

We use TPC-H Q1 [2] because it references only one relation, which makes it easier to prove intra-operator parallelism. The PN on NG_1 holds the entire Lineitem relation, whilst the PNs in NG_2 have a fifth of the original relation each ($\approx 1.97GB$).

Figure 8 depicts the GAP creating five LQAs and respective local queries. Each local query focuses on processing a section of the partitioned relation, whilst the PN in NG_1 must process the entire relation. By logic NG_2 should take five times less to process the query. The time taken to execute Q1 in NG_1 was 2258 seconds, whilst

[2] The variable parameters used to resolve Q1 are those suggested by the TPC-H manual.

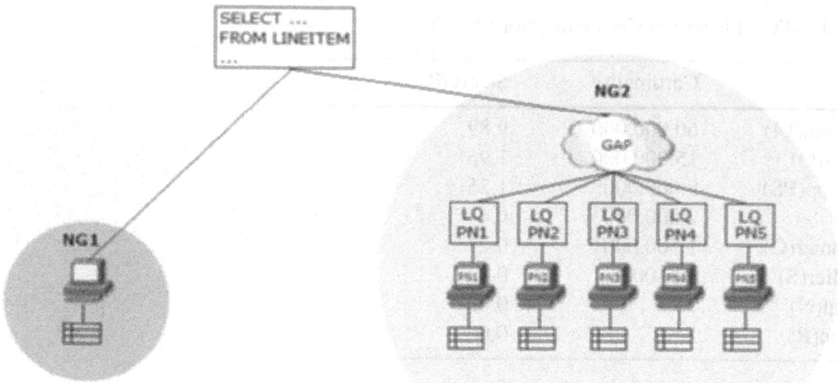
SELECT ...
FROM LINEITEM
...
NG2
GAP
LQ PN1
LQ PN2
LQ PN3
LQ PN4
LQ PN5
NG1

Fig. 8 Standalone versus Parallelization

the PNs of NG_2 took roughly 283 seconds each [3]. By using the GAP action plan C2 was able to obtain an execution ratio of approximately 0.13, which is higher than the predicted linear speedup (0.2) the query should obtain. One should note that the obtained super linearity is not directly associated with the GAP action plan, but with the saturation of the resources of the PN in NG_1, which ultimately lead it to execute the query more slowly. Yet it can be seen that by using the GAP action plan NG_2 was able to parallelize its processing, successfully promoting intra-operator parallelism.

Inter-operator Parallelization

In this experiment we have manually created a set of action plans that resolve query Q3 [4] of the TPC-H benchmark. Some of these plans use bushy trees to promote inter-operator parallelism. We show that the GAP is capable of fomenting bushy tree generation, and that it can differentiate between good and bad bushy plans.

Table 2 indicates how the three relations used in Q3 are partitioned amongst the PNs. It should be noted that it is not possible to directly join O with C, because they are partitioned on different attributes. To join both relations, O has to be repartitioned by its o_custkey attribute.

The action plans generated below are, to us, the most likely to process Q3 in the least amount of time. The border pointed blocks are datasets found in NG1, whilst the dashed bordered blocks are datasets found in NG_2. Repartitioning actions are represented by a dashed connecting line. The subscript characters are the attribute keys by which the datasets are partitioned, i.e. orderkey (ok) and custkey (ck). When

[3] One should note that the execution time presented for NG_2 also includes the merging of each PN's datasets into a single user readable dataset.

[4] The variable parameters used to resolve Q3 are those suggested by the TPC-H manual.

Table 2 Partition distribution per PN

Relation	Partition key	Partition size per PN (MB)
Lineitem (LI)	l_orderkey	2000
Orders (O)	o_orderkey	400
Customer (C)	c_custkey	55

two blocks from different NGs are at the same level then the plan is using a bushy tree.

The plans are illustrated in Figure 9.

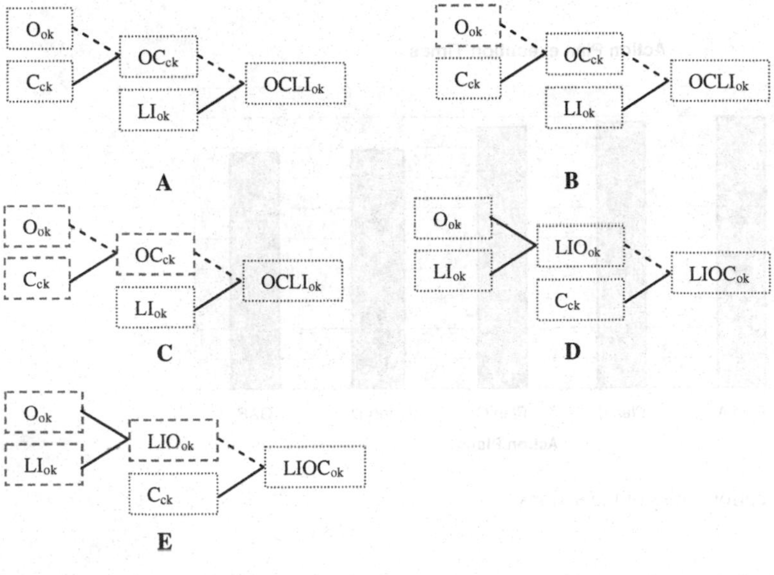
O_ok
C_ck
OC_ck
LI_ok
OCLI_ok
A
O_ok
C_ck
OC_ck
LI_ok
OCLI_ok
B
O_ok
C_ck
OC_ck
LI_ok
OCLI_ok
C
O_ok
LI_ok
LIO_ok
C_ck
LIOC_ok
D
O_ok
LI_ok
LIO_ok
C_ck
LIOC_ok
E

Fig. 9 Alternative execution plans considered for query Q3

The costs shown in Figure 10 are estimations given by the GAP for each of the abovementioned plans. Plans A, B and C are not chosen to be the best plan because their estimated costs are higher than those of plans D and E. Although the latter seem to present the same costs, E is 375 cost units smaller than D (not visible in the graph). We conclude that the GAP has attributed a smaller cost to E because it uses a bushy tree to optimize two independent actions, i.e. repartitioning of LIO_{ok} and the processing of C_{ck}. Plan D does no such thing and is therefore slower.

Figure 11 illustrates the real execution times of each proposed action plan. The graph confirms that plan E is slightly faster than D. Therefore the GAP was able to choose the fastest action plan as its best plan, and was also able to correctly convey the bushy tree to the DWPA middleware.

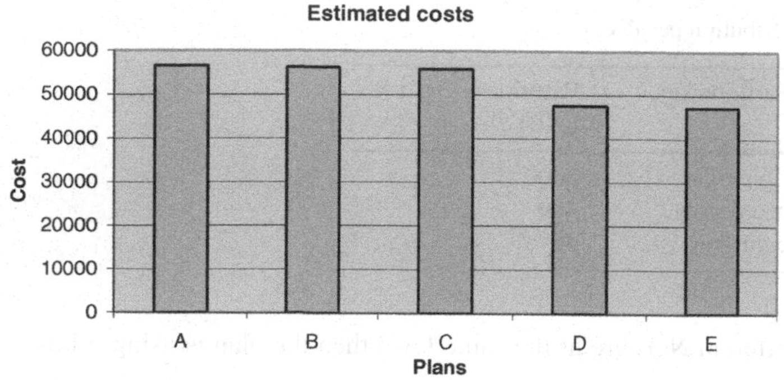
Estimated costs
60000
50000
40000
30000
20000
10000
0
Cost
A
B
C
D
E
Plans

Fig. 10 Estimated execution costs of the proposed action plans

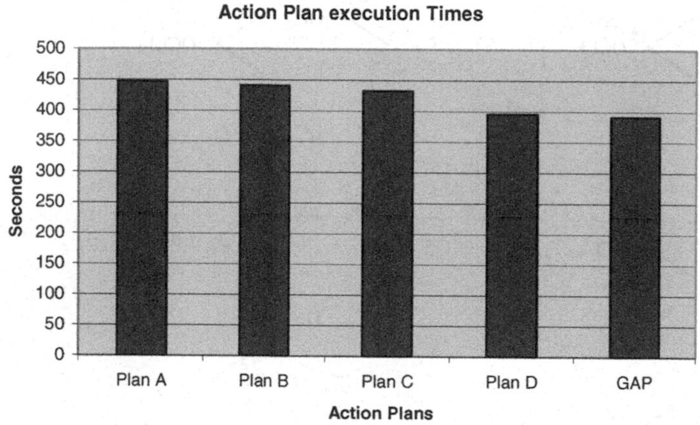
Action Plan execution Times
500
450
400
350
300
250
200
150
100
50
0
Seconds
Plan A
Plan B
Plan C
Plan D
GAP
Action Plans

Fig. 11 Execution times of the actions

Alternative Dataset Layouts

The manner by which relations are partitioned within a DW can influence the execution time of a query. If inadequately partitioned, the relations have to be repartitioned, which is always time consuming. A good practice is to have more than one partitioning strategy present in the DW.

In the next experiment we show that the GAP is aware of alternative partitioning layouts, and that it is capable of choosing which layout most benefits the response time of a specific query. To prove this, we partition the relations referenced in Figure 12 using three different partitioning strategies. The variable attribute shown in the figure, signifies that various dates will be allocated, thus generating various results.

The following three placement alternatives are used to populate a single NG containing five PNs.

```
SELECT c_custkey, c_name, sum(l_extendedprice*(1-l_discount)) as revenue, c_acctbal,
n_name

FROM customer, orders, lineitem, nation

WHERE c_custkey=o_custkey and l_orderkey=o_orderkey and o_orderdate>=variable and
o_orderdate<'1996-01-01' and l_returnflag='R' and c_nationkey=n_nationkey

GROUP BY c_custkey, c_name, c_acctbal, n_name

ORDER BY revenue desc;
```

Fig. 12 TPC-H query 10 with changes

- First - The Lineitem relation is partitioned throughout the PNs, whilst the remaining relations are replicated. The advantage of this strategy is that no repartitioning will ever be necessary.
- Second - Orders and Customer are equi-partitioned by their custkey attribute. Lineitem is partitioned by its orderkey attribute. When joining Orders to Lineitem a repartitioning action must be done beforehand.
- Third - Lineitem and Orders are equi-partitioned by the orderkey attribute. The Customer relation is replicated. No repartitioning is necessary to resolve this particular query.

Figure 13 shows the GAP estimated costs for each placement strategy. The first strategy is considered to be the slowest of all three alternatives. Although no repartitioning is needed to solve the query, the amount of data each PN has to join is very large in comparison to the other strategies. Join operations are very slow, PNs with little resources tend to become easily saturated. The second strategy is faster than the first because all relations are partitioned, thus the amount of data to join per PN is less than. However the partitioning strategy is not the most adequate for the query.

By observing Figure 13 one can see that the layout with the least cost estimation is the third layout. The GAP has attributed such low costs because it has realized

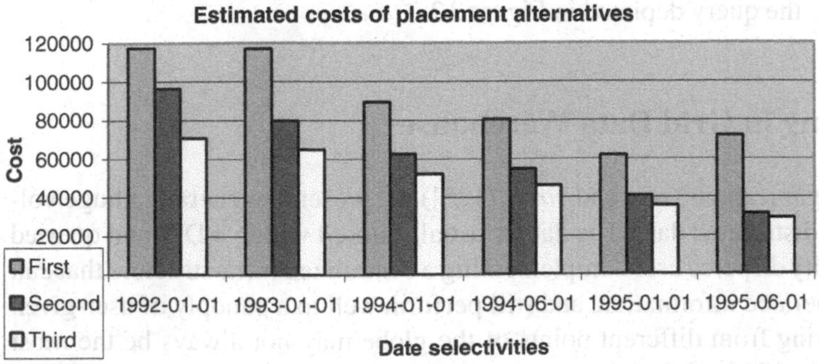
Estimated costs of placement alternatives
120000
100000
80000
60000
40000
20000
Cost
First
Second
Third
1992-01-01
1993-01-01
1994-01-01
1994-06-01
1995-01-01
1995-06-01
Date selectivities

Fig. 13 Estimated costs of placement alternatives

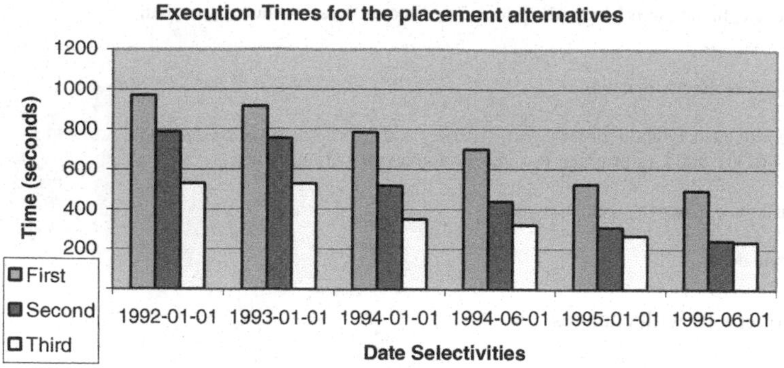
Execution Times for the placement alternatives
1200
1000
800
600
400
200
Time (seconds)
First
Second
Third
1992-01-01 1993-01-01 1994-01-01 1994-06-01 1995-01-01 1995-06-01
Date Selectivities

Fig. 14 Real execution times of placement alternatives

that no dataset repartitioning will be necessary for the query to be executed. It also knows that the largest datasets, i.e. Lineitem and Orders, have been partitioned to a fifth of their original size and that they are locally processable. The GAP estimates that the second strategy is worse than the third, because it has to repartition the intermediate results from the Orders/Customer join with the Lineitem relation. The estimated repartitioning actions are very costly when the selectivity of the Orders relation is relatively low (first three sets of bars). As the selectivity increases, it is estimated that less data will be exchanged between the PNs, due to the large progressive filtering of the intermediate Orders/Customer dataset (last three sets of bars). The layout schema with the least favorable results is the first strategy, although the layout avoids repartitioning, it cannot avoid the large processing times that are needed to join the replicated relations. The GAP knows that the Orders relation is five times larger than its partitioned counterparts found in the previous layouts, and that the Customer relation is also five times larger than the partitioned Customer relation found in the second layout. This ultimately forces the PNs to execute more data then would be necessary, if correctly partitioned.

Figure 14 illustrates the real execution times of the three partitioning strategies. The graph confirms the GAP estimation costs. The third strategy is the most adequate to execute the query depicted in Figure 12.

5 Scheduling in Grid Data Warehouses

Many global organizations (*real* and *virtual* [25]) are presently generating huge volumes of highly distributed data. The data is usually stored within a DW and queried by geographically disperse users. Implementing a centralized infra-structure that can store all the generated information, and still perform well to a panoply of user-given queries originating from different points of the globe may not always be the most appropriate choice. Therefore, it is important to extend the concept of the data warehouse to use the Grid as an infra-structure to interconnect the various data sources,

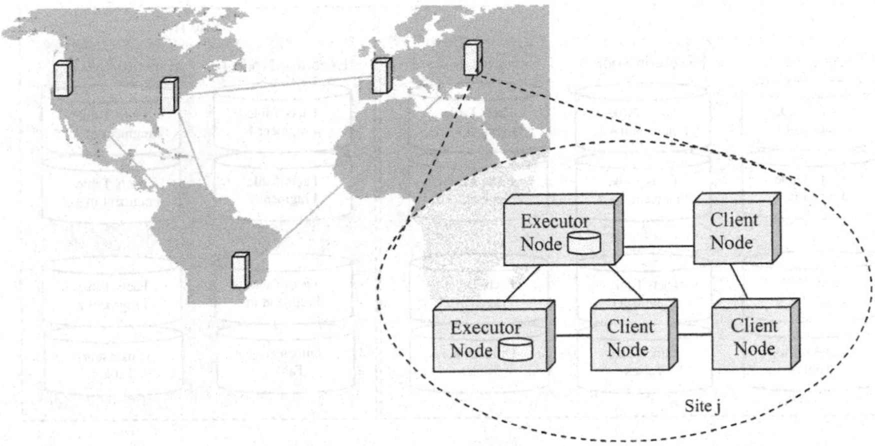
Executor
Node
Client
Node
Executor
Node
Client
Node
Client
Node
Site j

Fig. 15 Generic Grid-Data Warehouse Architecture

forming a distributed grid-based DW. Such a DW is able to achieve high performance levels at a reduced cost, when compared to centralized solutions. For instance, a grid-aware DW can distribute its workload across several sites, reducing hardware and network requirements.

We consider a generic grid-based DW to be formed by several (possibly highly distributed) sites, where each site has one or more nodes associated to it. Figure 15 illustrates the concept. All the sites may or may not belong to the same *real* organization.

Nodes can assume different roles. Some act as *client nodes*: nodes used by clients to submit queries to the DW. When submitted, a query scheduler assigns the queries to be executed in the nodes that possess the necessary data, i.e. *executor nodes*.

The hierarchical scheduling model is commonly used in grid-based systems. In such model, there exists a global community scheduler (or resource broker) and each site with executor nodes has a local scheduler. In Grid-DWPA we consider the use of the GAP module as a local optimizer (as discussed earlier, in Section 3) in order to make the parallel query scheduling engine-independent. But other local schedulers may be used to interact with the Grid-DWPA Community Scheduler.

When implementing grid-based DWs, two specific database-related factors must be well planned: (i) data placement and (ii) the job scheduling policy. We discuss both aspects in the following sub-sections.

5.1 Data Placement and Availability

The usage of a good data placement strategy can lead to faster response times and higher availabilities. To implement the DWPA data placement strategy, i.e. WBP, over a Grid infrastructure, one could place the partitioned data over different sites

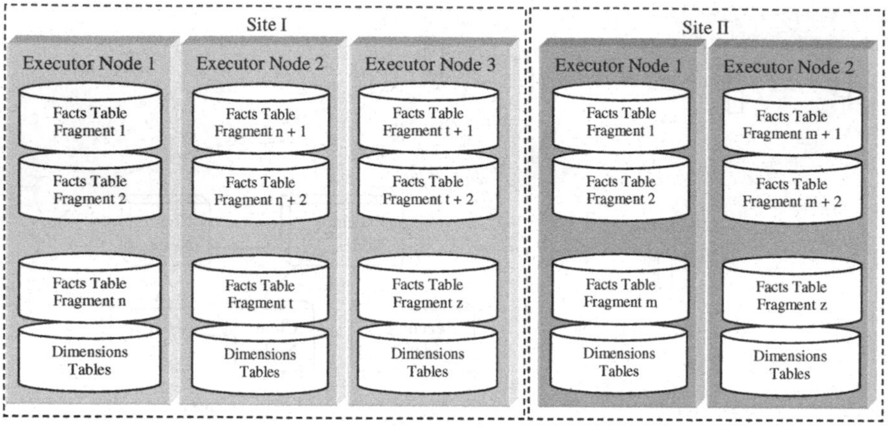
Site I
Site II
Executor Node 1
Executor Node 2
Executor Node 3
Executor Node 1
Executor Node 2
Facts Table Fragment 1
Facts Table Fragment 2
Facts Table Fragment n
Facts Table Fragment n + 1
Facts Table Fragment n + 2
Facts Table Fragment t
Facts Table Fragment t + 1
Facts Table Fragment t + 2
Facts Table Fragment z
Facts Table Fragment 1
Facts Table Fragment 2
Facts Table Fragment m
Facts Table Fragment m + 1
Facts Table Fragment m + 2
Facts Table Fragment z
Dimensions Tables

Fig. 16 Grid-DWPA Data Allocation Example

and replicate small tables at all sites. Although simple, this strategy can lead to some undesirable situations, specially related to availability.

One must take into consideration that such (virtual) organizations may be composed of several sites from distinct administrative domains, and that each domain can impose resource usage constraints, e.g. timeframe consumption or usage limits. These constraints may prohibit the access of remote users to local resources, including data, during certain periods. This leads to data unavailability during certain periods. The network infrastructure that composes the grid may also be unreliable, leading to situations where some sites may become temporarily inaccessible. Finally, the suggested placement strategy can also lead to high quantities of data interchange during query execution.

To resolve the limitations of abovementioned strategy the Grid-DWPA uses data replication, as in [29], to duplicate the partitioned relations to various sites.

In Grid-DWPA, some sites are selected to be processing sites. Such sites must have adequate storage capacity. Node availability, available processing power, data ownership and access patterns can also be used as site selection criteria. One can interpret a processing site as being a nodegroup (NG) (discussed earlier). Each NG has a replica of the small tables, all large tables are partitioned by the workload and the number of PNs that compose the NG. Figure 16 illustrates an example.

In Figure 16, two sites are represented by a total of five executor nodes. Each site can have a distinct number of processing nodes, yet the database schema is the same in all sites.

By using this data placement strategy, the system is able to achieve high levels of availability and throughput [11]. It can also be used in other situations where the system must deal with user-specified execution constraints (SLOs), as it is presented in the following subsections.

5.2 *Dynamic Query Scheduling in Grid-DWPA*

Besides data placement issues, another key aspect of any grid-based DW environment are the scheduling policies. Grid-DWPA uses specific scheduling policies to deal with throughput and user-defined SLOs. The Grid-DWPA Community Scheduler runs over a GRM system (e.g. Globus). The GRM system deals with general aspects such as authentication and efficient data transfer.

In Grid-based systems, users usually have the ability to establish service level related requirements, which are then used by the scheduler to make decisions. In fact, grid scheduling is satisfaction-oriented: scheduling is done in order to meet a users' expectations in terms of SLOs [52].

Time constraints, e.g. deadlines, are among the most commonly used SLO parameters in the Grid. In grid-based DWs, time constraints can be used to provide differentiation between interactive queries and report queries. For example, one can establish that interactive queries should be executed within a 20 second deadline, whilst report queries should be executed within 5 minutes.

The Grid-DWPA Scheduler implements a policy called Planned Scheduling in order to execute queries considering user-defined execution constraints. In Grid-DWPA , each GAP local scheduler estimates the resources that are needed to execute a task at its site. When considering deadlines as execution constraints, the GAP*TE* module of each processing site is used to estimate the time the local site would take to execute a task.

In some cases, the specified query deadline may not be achieved with the available resources. In such situations, two approaches can be considered:

- Specified deadlines are just objectives (soft deadlines): a query should be executed even though its deadline cannot be achieved;
- Specified deadlines are requirements (hard deadlines): if the deadline cannot be achieved, then the query does not have to be executed;

Planned Scheduling can be implemented with Soft or Hard Deadlines. The main steps of the planned scheduling strategy (illustrated in Figure 17) are:

1. The *Grid-DWPA Community Scheduler* is responsible for transforming the incoming query (job) into smaller parallelized queries (tasks) that are then executed over the DWPA-based DW (the transformation of a user-given query into a set of smaller parallelized queries is explained in [28]).
2. The generated tasks are placed into a *list of tasks* (LT) for execution. The list is then sent to the GAP optimizers at the available processing sites.
3. Each GAP Optimizer *i* transmits to the Query Scheduler a list (LT_i) containing the predicted execution time of each task.
4. The Community Scheduler verifies if the whole job is executable within a single site before its deadline is reached. If not, the scheduler tries to allocate the tasks in between sites, keeping in mind that it should involve the least amount of sites possible as well as choose an option with the least predicted execution time.

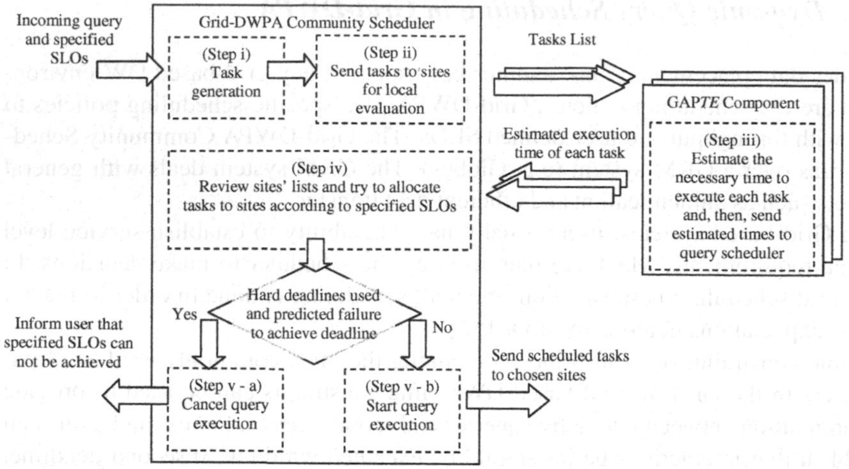
Incoming query and specified SLOs
Grid-DWPA Community Scheduler
(Step i) Task generation
(Step ii) Send tasks to sites for local evaluation
Tasks List
GAPTE Component
(Step iii) Estimate the necessary time to execute each task and, then, send estimated times to query scheduler
Estimated execution time of each task
(Step iv) Review sites' lists and try to allocate tasks to sites according to specified SLOs
Hard deadlines used and predicted failure to achieve deadline
Yes
No
Inform user that specified SLOs can not be achieved
(Step v - a) Cancel query execution
(Step v - b) Start query execution
Send scheduled tasks to chosen sites

Fig. 17 Planned Scheduling Main Steps

The objective of choosing the least amount of sites is to reduce network traffic by using Hierarchical Aggregation [28] (the results of the tasks executed by a certain site are merged and only the merged result is sent to a global node which will in turn merge the merged results into a final result).

5. If the Community Scheduler successfully allocates the tasks whilst fulfilling the specified execution time constraints or if soft deadlines are being used, then tasks are executed. Otherwise, job execution is canceled.

This strategy is in accordance to the Grid philosophy of site autonomy: each site has the ability to specify the necessary amount of time to execute a task, which may be done considering the existence of domain specific resource utilization constraints. With Planned Scheduling, when a site is unavailable, the Grid-DWPA Community Scheduler chooses another to execute the task, ultimately promoting data availability.

5.3 Experimental Evaluation

In this section we present several results demonstrating the usage of the Grid-DWPA Community Scheduler over a grid environment. The results include SLO-achievement rates, measured throughput and observed response times. The obtained values validate our main proposals and show that the Grid-DWPA Community Scheduler can lead to high levels of SLO achievement.

The test environment is composed of four heterogeneous executor nodes placed at different sites. We have used a 1GB TPC-H database with the Lineitem table partitioned into 30 fragments. The nodes have the SQL Server 2005 [45] database engine installed. Queries 1,5,7,8,9,10,14 and 18 from TPC-H constitute the test bed. Each query was used several times, considering different parameters, so as to create

Table 3 Experimental Environment Description

Site	Processor	RAM Memory
1	AMD Duron 1.6Ghz	752MB
2	AMD Athlon 1.5Ghz	480MB
3	AMD Athlon 1.5Ghz	736MB
4	AMD Duron 1.4Ghz	752MB
5	Intel Xeon Dual Processor 2.8Ghz	3.87GB

a workload of 50 queries. The various combinations of database and workload configurations lead to a total of almost 1500 tasks. The queries were submitted to the system according to several submission rates.

Table 3 briefly describes the most relevant hardware parameters of the computers used in the experiments. Site 1 is used by the Community Scheduler.

In our tests we consider four scheduling strategies: *Planned Scheduling with Soft Deadlines* (PSSD), *Planned Scheduling with Hard Deadlines* (PSHD), *Round-Robin Job Allocation*(RRJA) and *Round-Robin Task Allocation* (RRTA). In the *Round-Robin Job Allocation* each job is attributed to the sites in a round-robin fashion, however all tasks that constitute a job are allocated to the same site. In *Round-Robin Task Allocation*, the tasks of each job are allocated in a round-robin fashion. This strategy maximizes the parallelism of each job execution. Round-robin based job assignment strategies have been used before as a baseline for comparison in grid-scheduling papers [55] and also exists in GRM systems such as Legion [46].

First we present the mean throughput (Figure 18) and response times (Figure 19) for the compared strategies. Results show that RRJA is not a good choice in terms of throughput and response times. This strategy equally distributes jobs between heterogeneous sites, which leads to load imbalance. PSHD proves to have achieved the best results, the reasons for this are two: it is able to schedule tasks to be executed by the most appropriate nodes; the system executes fewer jobs than in other

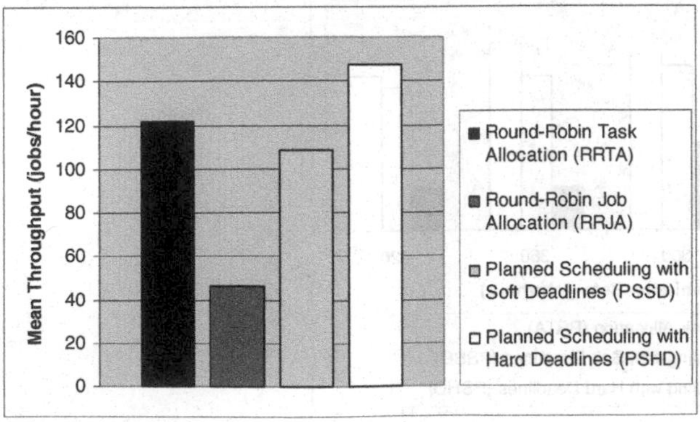
Mean Throughput (jobs/hour)
160
140
120
100
80
60
40
20
0
Round-Robin Task Allocation (RRTA)
Round-Robin Job Allocation (RRJA)
Planned Scheduling with Soft Deadlines (PSSD)
Planned Scheduling with Hard Deadlines (PSHD)

Fig. 18 Mean throughput - 240 jobs/hour submission rate

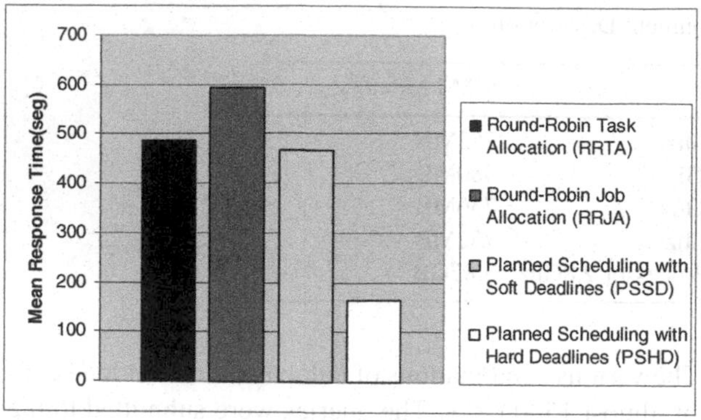
700
600
500
400
300
200
100
0
Mean Response Time(seg)
Round-Robin Task Allocation (RRTA)
Round-Robin Job Allocation (RRJA)
Planned Scheduling with Soft Deadlines (PSSD)
Planned Scheduling with Hard Deadlines (PSHD)

Fig. 19 Mean response time - 360 jobs/hour submission rate

policies, because it cancels jobs whose deadline cannot be achieved - this leads to more resource availability for the jobs that are effectively scheduled for execution.

When comparing RRJA and RRTA, we see that the latter presents smaller load imbalances. This can be justified by the fact that the difference between the sizes of tasks from the same job is smaller than the difference between sizes of different jobs. Hence, the imbalance of the total executed work using RRTA was smaller than the one of RRJA.

As discussed earlier, a main objective of grid scheduling is to achieve SLOs. In Figure 20, we present the SLO fulfillment rate (i.e. deadline achievement rate) for the tree scheduling strategies with the best throughput and response times. Results for different query submission rates are presented. The results obtained were much better when using Planned Scheduling methods than the round-robin based task allocation strategy.

Figure 20 shows that the use of hard deadlines has lead to better deadline achievement rates than those of soft deadlines. As discussed earlier, PSHD effectively

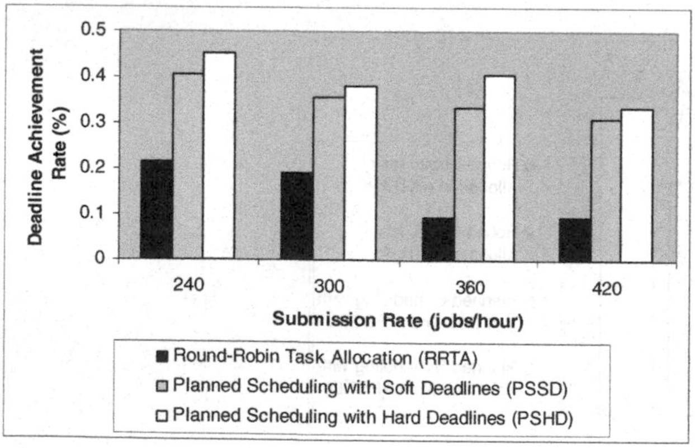
0.5
0.4
0.3
0.2
0.1
0
Deadline Achievement Rate (%)
240
300
360
420
Submission Rate (jobs/hour)
Round-Robin Task Allocation (RRTA)
Planned Scheduling with Soft Deadlines (PSSD)
Planned Scheduling with Hard Deadlines (PSHD)

Fig. 20 SLO fulfillment for several scheduling methods and submission rates

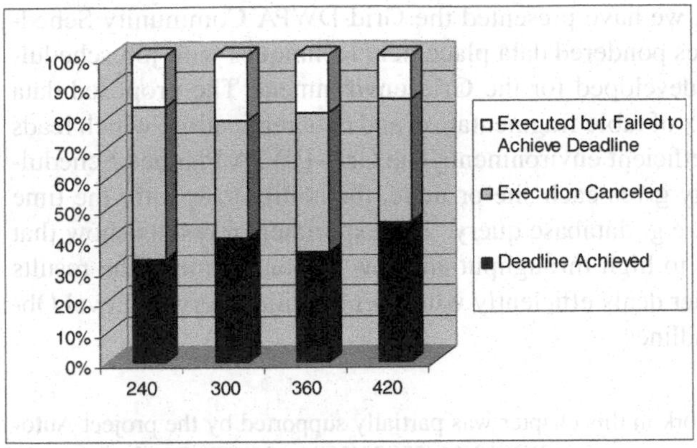
100%
90%
80%
70%
60%
50%
40%
30%
20%
10%
0%
240
300
360
420
Executed but Failed to Achieve Deadline
Execution Canceled
Deadline Achieved

Fig. 21 Execution rates of Planned Scheduling with Hard Deadlines

executes fewer jobs than PSSD. Figure 21 presents the rate of executed jobs against the total submitted jobs. It shows that a high number of queries were not executed, this is due to the small deadlines used. Another benefit of Planned Scheduling is that if one wants to use the execution deadline as a strict requirement, Planned Scheduling can immediately inform the user that the target deadline, in some situations, will not be achieved.

Hence, when dealing with strict deadlines, Planned Scheduling not only leads to high throughput and low response times, it also gives benefits to the user, by indicating in advance if the query is or is not executable within the predetermined deadline.

6 Conclusions

DWs store huge volumes of historical data for decision support purposes. To achieve high performance in such environments, special techniques and architectures have to be adopted. Our goal in the Data Warehouse Parallel Architecture (DWPA) is to offer a platform to run data warehouses in a parallel and distributed fashion in any context and with no specialized requirements. We have presented two inter-related modules of Grid-DWPA: the Global Action Planner - GAP, an engine-independent efficient DWPA parallel query optimizer, and the Grid-DWPA Community Scheduler, a module to schedule efficiently in a grid-distributed DWPA. We show experimentally that the GAP is effective in producing the best execution plans, being able to parallelize queries while using ordinary standalone database engines, which possess no parallel features. The optimization process is based on actions, i.e. communication actions (CA) and local query actions (LQA). Together they indicate how a user-given query should be processed within the DW. LQAs are an essential part of the optimization process, because they confer engine independence to our parallel optimizer.

For the Grid-DWPA infrastructure, it is also necessary to provide efficient solutions usable by global organizations. To provide efficient and transparent access to globally distributed DWs, we have presented the Grid-DWPA Community Scheduler. The strategy combines pondered data placement techniques with job scheduling strategies, especially developed for the Grid environment. The proposed data placement strategy is a mix of table fragmentation and data replication, which leads to a highly available and efficient environment. The Grid-DWPA Planned Scheduling job scheduling strategy gives each site or node, the ability to specify the time it needs to execute a task, e.g. database query. The experimental results show that Planned Scheduling leads to high throughput and low response times. The results also show that the scheduler deals efficiently with user specified Service Level Objectives, i.e. execution deadlines.

Acknowledgements. The work in this chapter was partially supported by the project Auto-DWPA: research and development project POSI/EIA/57974/2004 of FCT "Fundação para a Ciência e Tecnologia", Portugal, 2005-2008.

References

1. Abramson, D., Sosic, R., Giddy, J., Hall, B.: Nimrod: a tool for performing parametrised simulations using distributed workstations. In: HPDC 1995: Proceedings of the 4th IEEE International Symposium on High Performance Distributed Computing, p. 112. IEEE Computer Society, Washington (1995)
2. Alpdemir, N.M., Mukherjee, A., Gounaris, A., Paton, N.W., Watson, P., Fernandes, A.A., Smith, J.: Ogsa-dqp: A service-based distributed query processor for the grid. In: Proceedings of the Second UK e-Science All Hands Meeting (2003)
3. Babcock, B., Chaudhuri, S.: Towards a robust query optimizer: a principled and practical approach. In: SIGMOD 2005: Proceedings of the 2005 ACM SIGMOD international conference on Management of data, pp. 119–130. ACM, New York (2005), `http://doi.acm.org/10.1145/1066157.1066172`
4. Babu, S., Bizarro, P., DeWitt, D.: Proactive re-optimization. In: SIGMOD 2005: Proceedings of the 2005 ACM SIGMOD international conference on Management of data, pp. 107–118. ACM, New York (2005), `http://doi.acm.org/10.1145/1066157.1066171`
5. Baker, M., Buyya, R., Laforenza, D.: Grids and grid technologies for wide-area distributed computing. Softw. Pract. Exper. 32(15), 1437–1466 (2002), `http://dx.doi.org/10.1002/spe.488`
6. Ballinger, C., Fryer, R.: Born to be parallel: Why parallel origins give teradata an enduring performance edge. IEEE Data Eng. Bull. 20(2), 3–12 (1997)
7. Baralis, E., Paraboschi, S., Teniente, E.: Materialized views selection in a multidimensional database. In: Jarke, M., Carey, M.J., Dittrich, K.R., Lochovsky, F.H., Loucopoulos, P., Jeusfeld, M.A. (eds.) VLDB 1997, Proceedings of 23rd International Conference on Very Large Data Bases, Athens, Greece, August 25-29, pp. 156–165. Morgan Kaufmann, San Francisco (1997)
8. Baru, C., Fecteau, G.: An overview of db2 parallel edition. In: SIGMOD 1995: Proceedings of the 1995 ACM SIGMOD international conference on Management of data, pp. 460–462. ACM, New York (1995), `http://doi.acm.org/10.1145/223784.223876`

9. Bote-Lorenzo, M.L., Dimitriadis, Y.A., Gómez-Sánchez, E.: Grid characteristics and uses: A grid definition. In: Fernández Rivera, F., Bubak, M., Gómez Tato, A., Doallo, R. (eds.) Across Grids 2003. LNCS, vol. 2970, pp. 291–298. Springer, Heidelberg (2004)
10. Buyya, R., Abramson, D., Giddy, J.: Nimrod/g: An architecture for a resource management and scheduling system in a global computational grid. HPC 1, 283 (2000), http://doi.ieeecomputersociety.org/10.1109/HPC.2000.846563
11. de Carvalho Costa, R.L., Furtado, P.: Data warehouses in grids with high qos. In: Tjoa, A.M., Trujillo, J. (eds.) DaWaK 2006. LNCS, vol. 4081, pp. 207–217. Springer, Heidelberg (2006)
12. Chervenak, A.L., Palavalli, N., Bharathi, S., Kesselman, C., Schwartzkopf, R.: Performance and scalability of a replica location service. In: HPDC 2004: Proceedings of the 13th IEEE International Symposium on High Performance Distributed Computing, pp. 182–191. IEEE Computer Society, Washington (2004), http://dx.doi.org/10.1109/HPDC.2004.27
13. Chu, F., Halpern, J., Gehrke, J.: Least expected cost query optimization: what can we expect? In: PODS 2002: Proceedings of the twenty-first ACM SIGMOD-SIGACT-SIGART symposium on Principles of database systems, pp. 293–302. ACM, New York (2002), http://doi.acm.org/10.1145/543613.543651
14. Costa, M., Vieira, J., Bernardino, J., Furtado, P., Madeira, H.: A middle layer for distributed data warehouses using the dws-aqa technique. In: Pimentel, E., Brisaboa, N.R., Gómez, J. (eds.) JISBD, pp. 775–778 (2003)
15. Czajkowski, K., Foster, I.T., Karonis, N.T., Kesselman, C., Martin, S., Smith, W., Tuecke, S.: A resource management architecture for metacomputing systems. In: Feitelson, D.G., Rudolph, L. (eds.) IPPS-WS 1998, SPDP-WS 1998, and JSSPP 1998. LNCS, vol. 1459, pp. 62–82. Springer, Heidelberg (1998)
16. Deshpande, A., Ives, Z., Raman, V.: Adaptive query processing. Found. Trends databases 1(1), 1–140 (2007), http://dx.doi.org/10.1561/1900000001
17. DeWitt, D.J., Gray, J.: Parallel database systems: The future of high performance database systems. Commun. ACM 35(6), 85–98 (1992)
18. Evrendilek, C., Dogac, A.: Query decomposition, optimization and processing in multidatabase systems (1994), citeseer.ist.psu.edu/evrendilek94query.html
19. Fitzgerald, S.: Grid information services for distributed resource sharing. In: HPDC 2001: Proceedings of the 10th IEEE International Symposium on High Performance Distributed Computing, p. 181. IEEE Computer Society, Washington (2001)
20. Foster, I.: What is the grid? - a three point checklist. GRID today 1(6) (2002)
21. Foster, I., Kesselman, C.: Globus: A metacomputing infrastructure toolkit. The Internat. Journal of Supercomputer Applications and High Performance Computing 11(2), 115–128 (1997)
22. Foster, I., Kesselman, C.: The grid in a nutshell. Grid resource management: state of the art and future trends, 3–13 (2004)
23. Foster, I., Kesselman, C., Nick, J., Tuecke, S.: The physiology of the grid: An open grid services architecture for distributed systems integration. In: Globus Project Tech. Report (2002)
24. Foster, I., Kesselman, C., Tsudik, G., Tuecke, S.: A security architecture for computational grids. In: CCS 1998: Proceedings of the 5th ACM conference on Computer and communications security, pp. 83–92. ACM, New York (1998), http://doi.acm.org/10.1145/288090.288111

25. Foster, I., Kesselman, C., Tuecke, S.: The anatomy of the grid: Enabling scalable virtual organizations. Int. J. High Perform. Comput. Appl. 15(3), 200–222 (2001), http://dx.doi.org/10.1177/109434200101500302
26. Frey, J., Tannenbaum, T., Livny, M., Foster, I., Tuecke, S.: Condor-g: A computation management agent for multi-institutional grids. Cluster Computing 5(3), 237–246 (2002), http://dx.doi.org/10.1023/A:1015617019423
27. Furtado, P.: Workload-based placement and join processing in node-partitioned data warehouses. In: Kambayashi, Y., Mohania, M., Wöß, W. (eds.) DaWaK 2004. LNCS, vol. 3181, pp. 38–47. Springer, Heidelberg (2004)
28. Furtado, P.: Hierarchical aggregation in networked data management. In: Cunha, J.C., Medeiros, P.D. (eds.) Euro-Par 2005. LNCS, vol. 3648, pp. 360–369. Springer, Heidelberg (2005)
29. Furtado, P.: Replication in node partitioned data warehouses. In: VLDB Workshop on Design, Implementation, and Deployment of Database Replication (DIDDR) (2005)
30. Ganguly, S.: Design and analysis of parametric query optimization algorithms. In: VLDB 1998: Proceedings of the 24th International Conference on Very Large Data Bases, pp. 228–238. Morgan Kaufmann Publishers Inc, San Francisco (1998)
31. Gounaris, A., Smith, J., Paton, N.W., Sakellariou, R., Fernandes, A.A.A., Watson, P.: Adapting to changing resource performance in grid query processing. In: Pierson, J.-M. (ed.) VLDB DMG 2005. LNCS, vol. 3836, pp. 30–44. Springer, Heidelberg (2006)
32. Grimshaw, A.S., Wulf, W.A., Team, C.T.L.: The legion vision of a worldwide virtual computer. Commun. ACM 40(1), 39–45 (1997)
33. Hasan, W.: Optimization of sql queries for parallel machines. Ph.D. thesis, Stanford University, Stanford, CA, USA (1996)
34. Hasan, W., Motwani, R.: Coloring away communication in parallel query optimization. In: VLDB 1995: Proceedings of the 21st International Conference on Very Large Data Bases, pp. 239–250. Morgan Kaufmann Publishers Inc., San Francisco (1995)
35. Hillson, S., Hobbs, L., Lawande, S.: Improve results with query rewrite (2008), http://www.oracle.com/technology/oramag/oracle/03-sep/o53business.html (last visited, April 2008)
36. Hong, W., Stonebraker, M.: Optimization of parallel query execution plans in xprs. Distrib. Parallel Databases 1(1), 9–32 (1993), http://dx.doi.org/10.1007/BF01277518
37. HP: Hp neoview parallel query optimizer, http://whitepapers.techrepublic.com.com/whitepaper.aspx?docid%=283608 (last visited, April 2008)
38. Hulgeri, A., Sudarshan, S.: Anipqo: almost non-intrusive parametric query optimization for nonlinear cost functions. In: VLDB 2003: Proceedings of the 29th international conference on Very large data bases, pp. 766–777. VLDB Endowment (2003)
39. Ioannidis, Y.E., Ng, R.T., Shim, K., Sellis, T.K.: Parametric query optimization. VLDB J. 6(2), 132–151 (1997)
40. Kossmann, D., Stocker, K.: Iterative dynamic programming: a new class of query optimization algorithms. ACM Trans. Database Syst. 25(1), 43–82 (2000), http://doi.acm.org/10.1145/352958.352982
41. Krauter, K., Buyya, R., Maheswaran, M.: A taxonomy and survey of grid resource management systems for distributed computing. Softw. Pract. Exper. 32(2), 135–164 (2002)
42. Kruskal, J.: On the shortest spanning subtree of a graph and the traveling salesman problem. Proceedings of the American Mathematical Society 7(1), 48–50 (1956)
43. Lawrence, M., Rau-Chaplin, A.: The olap-enabled grid: Model and query processing algorithms. In: HPCS 2006: Proceedings of the 20th International Symposium on High-Performance Computing in an Advanced Collaborative Environment (2006)

44. Lohman, G.M., Mohan, C., Haas, L.M., Daniels, D., Lindsay, B.G., Selinger, P.G., Wilms, P.F.: Query processing in r*. In: Query Processing in Database Systems, pp. 31–47. Springer, Heidelberg (1985)
45. Microsoft: Microsoft sql server 2005 home page (2008), http://www.microsoft.com/sql/ (last visited, April 2008)
46. Natrajan, A., Humphrey, M.A., Grimshaw, A.S.: Grid resource management in legion. Grid resource management: state of the art and future trends, 145–160 (2004)
47. O'Neil, P., Graefe, G.: Multi-table joins through bitmapped join indices. SIGMOD Rec. 24(3), 8–11 (1995), http://doi.acm.org/10.1145/211990.212001
48. O'Neil, P.E., Quass, D.: Improved query performance with variant indexes. In: Peckham, J. (ed.) SIGMOD 1997, Proceedings ACM SIGMOD International Conference on Management of Data, Tucson, Arizona, USA, May 13-15, pp. 38–49. ACM Press, New York (1997)
49. Oracle: Oracle real application clusters (2008), http://www.oracle.com/technology/products/database/clustering%/index.html (last visited, April 2008)
50. Prim, R.C.: Shortest connection networks and some generalizations. The Bell System Technical Journal 3, 1389–1401 (1957)
51. Ranganathan, K., Foster, I.: Computation scheduling and data replication algorithms for data grids. Grid resource management: state of the art and future trends, 359–373 (2004)
52. Roy, A., Sander, V.: Gara: a uniform quality of service architecture. Grid resource management: state of the art and future trends, 377–394 (2004)
53. Selinger, P.G., Astrahan, M.M., Chamberlin, D.D., Lorie, R.A., Price, T.G.: Access path selection in a relational database management system. In: SIGMOD 1979: Proceedings of the 1979 ACM SIGMOD international conference on Management of data, pp. 23–34. ACM, New York (1979), http://doi.acm.org/10.1145/582095.582099
54. Shasha, D., Wang, T.L.: Optimizing equijoin queries in distributed databases where relations are hash partitioned. ACM Trans. Database Syst. 16(2), 279–308 (1991), http://doi.acm.org/10.1145/114325.103713
55. Silaghi, G.C., Arenas, A.E., Silva, L.M.: A utility-based reputation model for service-oriented computing. In: Priol, T., Vanneschi, M. (eds.) Toward Next Generation Grids. CoreGRID Series, pp. 63–72. Springer, Heidelberg (2007)
56. Smith, J., Gounaris, A., Watson, P., Paton, N.W., Fernandes, A.A.A., Sakellariou, R.: Distributed query processing on the grid. In: Parashar, M. (ed.) GRID 2002. LNCS, vol. 2536, pp. 279–290. Springer, Heidelberg (2002)
57. Tannenbaum, T., Wright, D., Miller, K., Livny, M.: Condor – a distributed job scheduler. In: Beowulf Cluster Computing with Linux, MIT Press, Cambridge (2001)
58. Thain, D., Tannenbaum, T., Livny, M.: Condor and the grid. In: Grid Computing: Making the Global Infrastructure a Reality. John Wiley & Sons Inc., Chichester (2003)
59. TPC: Transaction processing performance council (2008), http://www.tpc.org/ (last visited, April 2008)
60. Venugopal, S., Buyya, R.: A deadline and budget constrained scheduling algorithm for escience applications on data grids. In: Hobbs, M., Goscinski, A.M., Zhou, W. (eds.) ICA3PP 2005. LNCS, vol. 3719, pp. 60–72. Springer, Heidelberg (2005)
61. Wehrle, P., Miquel, M., Tchounikine, A.: A grid services-oriented architecture for efficient operation of distributed data warehouses on globus. In: AINA 2007: Proceedings of the 21st International Conference on Advanced Networking and Applications, pp. 994–999. IEEE Computer Society, Washington (2007), http://dx.doi.org/10.1109/AINA.2007.13

Database Support for Automatic Web Queries Categorization

Ernestina Menasalvas Ruiz and Santiago Eibe Garcia

Abstract. The increasing usage of web search engines together with the potential added value of knowing user interests when submitting a query are in the roots of the categorization of web queries research. Categorizing queries is challenging both for the problems associated to gathering and analyzing user context information and for the ones related to deployment of the knowledge obtained. Related to the first one, an interesting open problem is to analyze the mapping, if any, between user queries and the content shown by the portal at the main page. The automatization of this problem would be very beneficial and among the challenges we underlay the implementation of a database to support the process. In this chapter, we firstly review the main approaches for web query categorization and then we concentrate on analysing the process and the database support required for its automatization.

Keywords: Query Categorization, data warehouse, search engines, data mining.

1 Introduction

One of the challenges for search engines is to be able to understand user queries and respond according to them. From the portal perspective three dimensions interact with each other: the user intentions should be reflected in the information that appears on the site and this should be embedded in the portal structure that is also changing according to the published content and user information needs. Consequently, one of the challenges, is to find a data analysis model that merges information from all three perspectives: the user intentions, the system structure and the content of the site. On the other hand, IR researchers have begun to expand their efforts to understand the nature of the information need that users express in web search queries. All of these concerns fall into the general area of query understanding [3]. The central idea is that there is more information present in a user query than simply the topic of focus, and that harnessing this information can lead to the

Ernestina Menasalvas Ruiz and Santiago Eibe Garcia
Facultad de Informatica UPM
e-mail: {emensalvas, seibe}@fi.upm.es

D. Zakrzewska et al. (Eds.): Meth. and Support. Tech. for Data Analys., SCI 225, pp. 57–70.
springerlink.com

development of more effective and efficient information retrieval systems [14, 30] than in turn will aid to improve content and structure of the sites.

However, there are several issues [4, 15] that make analysis of web search queries very difficult:

1. The web is a dynamic collection: its data, users and popular queries are constantly changing
2. Web queries are typically very short
3. Web query terms can be noisy
4. One user query often has multiple meanings and may also evolve over time

These issues complicates manual categorization of queries (for study, or to provide training data) and poses great challenges for automated query classification [12, 18]

Consequently, the results obtained so far are not completely satisfactory being the main drawbacks not only a rapid changing technology but also the context analysis and storage.

Problems associated to measuring the context so far [1] have turn research into analysis of the search string alone. But interpreting intention using words alone is known not to be the best approach because of polisemies, semantics of language and of course context. To deal with this problem methods based on ontologies and meta-searches have been proposed [24, 28]. On the other hand, topic and event evolution analysis aiming at trend detection and tracking (TDT) has considerably have gained in interest during the last years. Consolidated studies [20] have concentrated on identifying and visualizing dynamically evolving text patterns from news data streams. Detecting and understanding user behaviour and relating user intentions to emerging topic trends in data streams still continues to remain a huge challenge for making search engines in "real-time" responsive to user's information needs.

In this paper we will review the process of web categorization and will analyze the steps to systematization of the process. Related to this aspect we will highlight the importance of the database support and will present an approach based on a data warehouse so to categorize queries related to their mapping with topic presented in the site main page.

The rest of the paper is organized as follows. To start with, in section 2 a review of previous studies on categorization of queries is found. In section 3.1 the presented approach is presented, firstly we define the basics of the method and later the stages of the process and the design of the data-warehouse that supports all the process. In section 3.4 the design of a database to support the process explained is shown. To conclude, section 4 presents the outlook and future of the research in the field.

2 Related Work

2.1 Preliminaries

The benefits of the categorization on improving effectiveness highly depends on whether classification is performed before or after the query is used to retrieve documents. In any case, effectiveness in the retrieval process is not the only benefit of categorization, in fact the main benefit is knowing the user intention and consequently sponsors of the site can take advantage of this knowledge. Furthermore, in Web search, many search engine companies are interested in providing commercial services in response to user queries, including targeted advertisement, product reviews, and valued added services such as banking and transportation, according to the categories. Therefore, some previous work focuses on classifying search results, instead of classifying queries directly, as an alternative way to understand queries [10].

Query classification can be used to effectively organize results. So, an important application in Web search of post-retrieval query classification is to organize the large number of Web pages in the search result after the user issues a query, according to the potential categories of the results. In [4] the pre versus post-retrieval classification effectiveness and the effect of training explicitly from classified queries versus bridging a classifier trained using a document taxonomy is examined.

In [5] authors examine three approaches to topical categorization of general web queries: matching against a list of manually labeled queries, supervised learning of classifiers and mining of selectional preference rules from large unlabeled query logs. Each approach has its advantages in tackling the web query classification recall problem, and combining the three techniques allows to classify a substantially larger proportion of queries than any of the individual techniques.

In any case, the problem of how to automatically and effectively classify a large number of queries that are inherently short, noisy, and ambiguous remains. An alternative approach is through some query expansion mechanism before the query is used to retrieve documents.

In [18] categories are obtained through clustering query set or other unsupervised learning methods. In clustering web queries, the problem is no longer lack of queries, but lack of features in any individual query. This is arguably an even greater problem for unsupervised than supervised approaches, which can at least latch on to individual features correlated with a class. Some query clustering methods have attacked this problem by clustering session data containing multiple queries and click-through information from a single user interaction [2].

Other interesting work has emerged on using query logs. In [2, 31] the authors propose to expand the meanings of user queries mapping from queries to clicked Web pages. However, it does not automatically provide the target categories because the pages themselves still need to be mapped to different categories. In addition, navigation logs are not always available for the timely training and application of query classification models.

In [7] a method that exploits page counts and text snippets returned by the search engine are used to define different scores to compute similarity of words. These scores are integrated later with support vector machines leading a robust similarity measurement that improves significantly the F-measure in entity disambiguation tasks and in community mining task.

In [22] an approach to add useful meta data to search results by fast-feature techniques is explored. The main motivation of the paper is to stress the importance of lightweight rapid techniques for categorizing search results into meaningful and stable categories. The approach presented in this paper also deals with a fast-feature categorization method but instead of being based on features of the results we based our approach on features of the submitted queries.

In [26] the author presents an approach to classifying large quantities of search queries automatically called *query enrichment*, which takes a short query and classifies it into target categories by making use of a set of intermediate objects. The process consists of two major steps: first, replace the query by a set of objects wherein the meanings of the query are embedded; secondly, classify the query based on the set of objects into ranked target categories. The authors propose to use the results from search engines [27] to provide the intermediate objects that are Web pages and category taxonomies such as that of the Open Directory Project (ODP). Classification requires training data on line for each category that are founded in the target taxonomy.

2.2 Query Understanding

Understanding information needs in user web search as has been already motivated is attracting a lot of attention. Several studies have attempted to classify user queries in terms of the informational actions of users. In early work [8] manually classified a small set of queries into transactional, navigational, and informational tasks using a pop-up survey of Alta Vista users, and manual inspection. In [25] a framework for understanding the underlying goals of user searches and their experience in using that framework to manually classify queries from a web search engine is presented. More recently, [13] analyze samples of queries and identify three broad classifications of user intent as expressed by their query: informational, navigational, and transactional. Their study showed show that more than 80% of Web queries are informational and only 20% are navigational or transactional. However, most of those approaches do not consider the searching behavior of the end users. A wide spectrum of research is on its way to personalize the web search results to meet the user needs. In [17] authors propose an algorithm to group user queries semantically to form what they refer to as "super concepts". Each set is composed of related queries and is organized in a hierarchical structure to extract the general user search patterns and improve retrieval effectiveness by utilizing the generic behavioral patterns.

In [29] the ambiguity of queries is analyzed. First, the authors construct the taxonomy of query ambiguity, and ask human annotators to manually classify queries based upon it. From manually labeled results, they find that query ambiguity is to some extent predictable and use a supervised learning approach to automatically classify queries as being ambiguous or not resulting in a 16% approximately of queries in a real search log being ambiguous in the experimental results.

By surveying the literature, the authors in the same paper summarize the following three types of queries from being ambiguous to specific.

- Type A (Ambiguous Query): a query that has more than one meaning;
- Type B (Broad Query): a query that covers a variety of subtopics and a user might look for one of the subtopics by issuing another query.
- Type C (Clear Query): a query that has a specific meaning and covers a narrow topic.

For this reason it becomes very important not only to analyze the terms in the query but to go deeper analysing further relationships of these terms. Therefore,in [13] the authors examine other aspects of the interaction including:

- number of query reformulations
- selection of vertical, use of system feedback, and result page
- uses question words;
- queries with natural language terms;
- queries containing informational terms (e.g. list, play list, etc.);
- queries that were beyond the first query submitted;
- queries where the searcher viewed multiple results pages;
- queries length (i.e., number of terms in a query) greater than 2; and
- queries that do not meet criteria for navigational or transactional.

In fact analyzing more information related to the query than the query itself makes it possible to better understand the user intention. Nevertheless, few studies try to map the query with the contents shown in a particular moment in the portal what would make it possible to analyze the way in which the site respond to the user needs and whether the user is influenced by the contents presented in the site.

On the other hand, in [23] the authors present an approach to automate the process of user goal identification by proposing two types of features for the goal-identification task: user-click behavior and anchor-link distribution. The experimental evaluation shows that by combining these features the goals for 90% of the queries studied can be successfully identified.

2.3 Temporal Profiles of Queries

Another important aspect to be analyzed is the temporality of queries. In [16] this problem is analyzed showing how documents with timestamps, such as email and news, can be placed along a time line. The time line for a set of documents returned in response to a query gives an indication of how documents relevant to that query are distributed in time. The authors explain how by examining the time line of a query result set makes it possible to characterize both how temporally dependent the topic is, as well as how relevant the results are likely to be. The authors also show that properties of the query result set time line can help predict the mean average precision of a query. These results show once again that meta-features associated with a query can be combined with text retrieval techniques to improve our understanding and treatment of text search on documents with timestamps.

On the other hand, the structure of the most visited regions of the Web is altered at the timescale from hours to days. In [21] authors analysed continuous search queries, search queries over a period of time. Authors defend that the improvement of continuous search queries may concern not only the quality of retrieved results but also the freshness of results. In some cases a user should be notified immediately since the value of the respective information decreases quickly, as e.g. news about companies that affect the value of respective stocks, or sales offers for products that may no longer be available after a short period of time. This is particularly shown in news sites. In [19] the authors analyze the dynamics of visits of a major news portal, representing the prototype for such a rapidly evolving network. In [9] a topic mining framework that supports the identification of meaningful topics from news stream data is proposed. The approach presented is based on retrieving News articles and applying data mining to produce patterns that are stored in a data base. The clustering technique proposed is incremental so to deal with the high rate of documents update.

We can conclude this section by underlying the paramount influence that understanding queries has to categorize them and how the understanding of the query is not only based on the terms of the query but on context elements such as temporality and it relationship with contents of the site in which the query is submitted.

3 Database Support for Web Query Categorization

3.1 The Process of Categorization

In [11] a method to categorize the queries submitted to a search engine searching for results across the corporate content of a company was presented. The method is based upon two assumptions: i) the taxonomy of concepts can be extracted from the site structure, ii) as in [6] no external sources are used for categorization but the queries themselves and the information of the site structure.

We review in what follows the proposed method stressing the database support required in each of the stages of the process. In fact the method is composed of the following stages (see figures 1, 2 and 3 for details)

1. Taxonomies creator. This process is in charge of analizing the web site structure so to create a concept hierarchy based on the informacion of the site directory tree. The information gathered by this process will be later used to map queries with site structure. Consequently, this information has to be stored so that to be used later on the classification process.
2. Web Log cleaning: This is the tipical process in web mining in charge of cleaning and prepocessing the requests sent to the search engine so to be used by the mining process. All the information obtained as a result has also to be properly stored to be accesible for the mining processes.
3. Events identification. We have used the term event to name any information (piece of news, video, link, ...) that appears on the main page in each particular moment. One of the main goals of the approach is to find mappings between

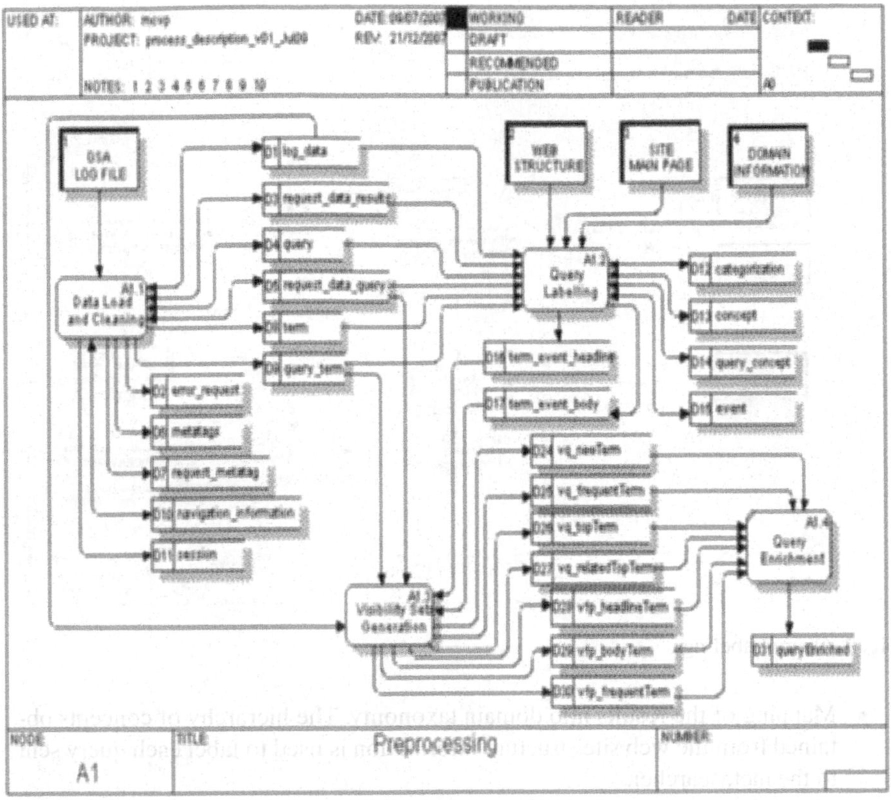
USED AT:
AUTHOR:
PROJECT:
NOTES: 1 2 3 4 5 6 7 8 9 10
WORKING
DRAFT
RECOMMENDED
PUBLICATION
READER
DATE
CONTEXT:
A0
LOG FILE
D1 log_data
D3 request_data_results
D4 query
D5 request_data_query
D8 term
D9 query_term
A1.1 Data Load and Cleaning
D2 error_request
D6 metatags
D7 request_metatag
D10 navigation_information
D11 session
WEB STRUCTURE
SITE MAIN PAGE
DOMAIN INFORMATION
A1.2 Query Labelling
D12 categorization
D13 concept
D14 query_concept
D15 event
D16 term_event_heading
D17 term_event_body
D24 vq_newTerm
D25 vq_frequentTerm
D26 vq_topTerm
D27 vq_relatedTopTerms
D28 vfp_headlineTerm
D29 vfp_bodyTerm
D30 vfp_frequentTerm
A1.3 Visibility Sets Generation
A1.4 Query Enrichment
D31 queryEnriched
NODE
A1
TITLE
Preprocessing
NUMBER

Fig. 1 Preprocessing and Enrichment Activities

the user queries and the content shown in the main page. Consequetly all the informatior related to events is stored for later usage.

4. Enrichment of queries with Visibility measurements. The approach we propose to classify queries tries to find relationship among queries along time and between queries and information being shown at the main page. In order to establish such relationship we have define some measurements called visibility measurements that in short intend to capture how visible one term in one query is in other queries being submitted to the engine and how visible is one term on the contents the user gets on the main page. Once again the information related to visibility of each query is stored in the supporting database.
5. Manual labeling of queries (site structure) In order to obtain a classification model a set o queries previously classified is required. To obtain such a training set the following process is performed:
   - Search sets of results for a selected set of queries. A set of selected queries is sent to a metasearcher and the results obtained are stored.

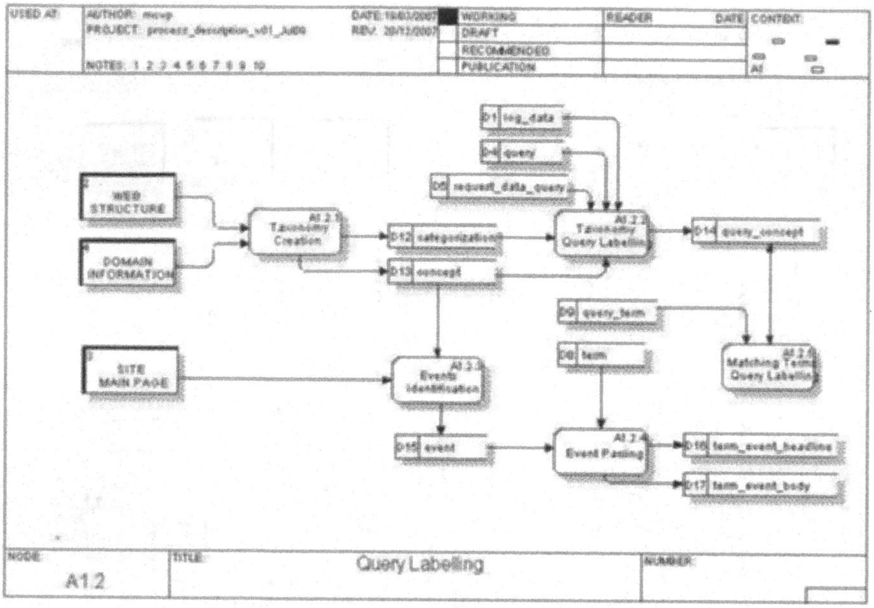
USED AT:
AUTHOR:
PROJECT:
NOTES: 1 2 3 4 5 6 7 8 9 10
WORKING
DRAFT
RECOMMENDED
PUBLICATION
READER
DATE
CONTEXT:
WEB STRUCTURE
DOMAIN INFORMATION
SITE MAIN PAGE
Taxonomy Creation
Taxonomy Query Labelling
Events Identification
Event Parsing
Matching Term Query Labelling
D1 log_data
D4 query
D5 request_data_query
D12 categorization
D13 concept
D14 query_concept
D9 query_term
D8 term
D15 event
D16 term_event_headline
D17 term_event_body
NODE:
A1.2
TITLE:
Query Labelling
NUMBER:

Fig. 2 Query Labeling

- Mapping of the results into domain taxonomy. The hierarchy of concepts obtained from the web site structure information is used to label each query sent to the metasearcher.

6. Front-page labeling of queries (figure 2). According to the information obtained of each request and in particular taking the query content into account the request are labelled as related or not to the events on the main page. Labelling does not mean adding one lables but to the contrary, labels are added to take different time slot into account.
7. Mining process. Using the enriching information of the database, the mining procedures will obtain patterns on the submitted queries. The process of mining in short is as follows:

   - Cluster queries. According to different criteria mainly based on visibility queries are clustered
   - Labelling of queries according to clustering results
   - Clasiffication of queries based on previous results and content information

8. Evaluation

In what follows we detail some of the important aspects of the process putting special attention to the taxonomies of concepts, visibility of words in queries and the database support.

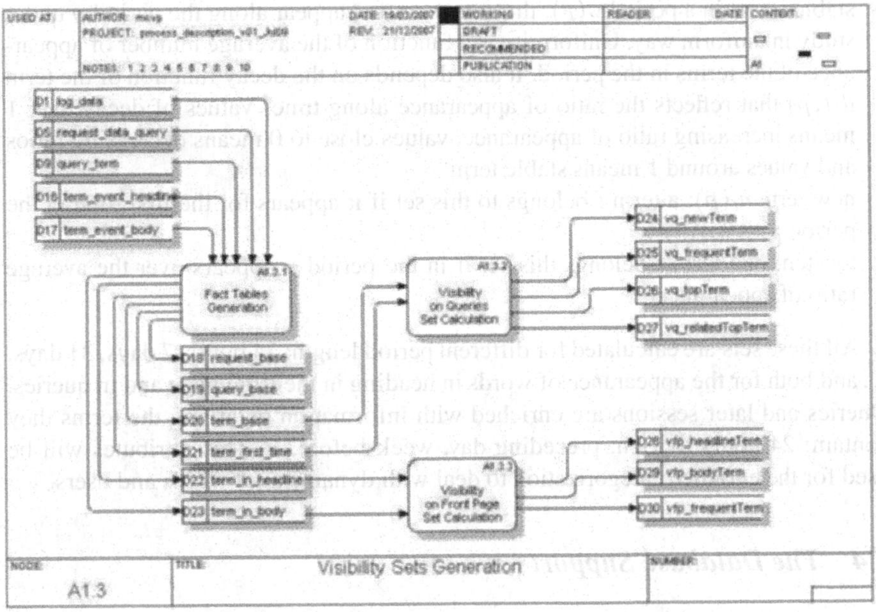
USED AT:
AUTHOR:
PROJECT:
NOTES: 1 2 3 4 5 6 7 8 9 10
WORKING
DRAFT
RECOMMENDED
PUBLICATION
READER
DATE
CONTEXT:
D1 log_data
D5 request_data_query
D9 query_term
D16 term_event_headline
D17 term_event_body
Fact Tables Generation
D18 request_base
D19 query_base
D20 term_base
D21 term_first_time
D22 term_in_headline
D23 term_in_body
Visibility on Queries Set Calculation
D24 vq_newTerm
D25 vq_frequentTerm
D26 vq_topTerm
D27 vq_relatedTopTerm
Visibility on Front Page Set Calculation
D28 vfp_headlineTerm
D29 vfp_bodyTerm
D30 vfp_frequentTerm
NODE:
A1.3
TITLE:
Visibility Sets Generation
NUMBER:

Fig. 3 Visibility Sets Generation

3.2 Taxonomies of Concepts

We propose to deal with two kind of categories:

- The one inherent in the site structure: this is very stable taxonomy as the concepts and structure of the site does not change very often
- The one extracted from the front page: this is a very changing taxonomy but it reveals interesting information regarding the dynamics of the users and their interests

3.3 Visibility of Words in Queries

Interestingness of the query highly depends on features of the terms contained in the query. Consequently we firstly categorize terms and then according to them we label queries. Two criteria are taken into account:

- visibility of the term on the queries
- visibility of the term on the front page (see figure 3)

Consequently we define the *visibility* of a term in a period of time $V(t,p)$ based on its appearance frequency (number of times it appears) in the period. According to this coefficient we define:

- stable terms in a period $s_t(p)$: those terms t that appear along the period p under study in uniform way. Uniformity is a function of the average number of appearance of the terms in the period. It also depends on the decay function of the term $d(t,p)$ that reflects the ratio of appearance along time. Values of decay over 1 means increasing ratio of appearance, values close to 0 means decreasing ratios and values around 1 means stable term
- new term $n_t(p)$: a term t belongs to this set if it appears for the first time in the period p
- top-ten: one term t belongs this set if in the period p appears over the average ratio of appearance

All these sets are calculated for different period length: 24 hours, 7 days, 31 days, ...and both for the appearance of words in heading in the front page and in queries. Queries and later sessions are enriched with information regarding the terms they contain: 24 hours top ten, preceding day, week before, ...This attributes will be used for the enriched categorization to deal with dynamism of words and users.

3.4 The Database Support

Data to be dealt with for web query categorization are generated as streams of data: i) queries are generated in a stream fashion reflecting the behavior of a changing population of users, ii) contents presented in the main page are being changed in a continuous fashion to try to respond to users needs as well as iii) the site structure also changes so to accommodate to these needs. Consequently, the process of mining has to be executed so to deal with these streams of data.

Observe that fact tables to store the information of the appearance of terms in the headings or in the body of text are needed. On the other hand, the labeling of each query in each time stamp established is also required in order to be able to calculate the aggregated data for the classification stream mining algorithms. Although this design has been performed for a particular news site, observe that tables and concepts are general enough to be applied in any other domain.

Due to the large volume of data generated in this process and in order to avoid the user monitoring the preprocessing stage the process of extraction, cleaning and loading of the information into the mining supporting database has been also automated. The main support for automating the process in our case is based on metadata being generated and stored so that the preprocessing functions will query this information and update it when needed.

For example, one of the main points prior to mining (classification mining) has to do with all the process of event interpretation and visibility of measurements calculation. This process is performed in an autonomous way thanks to the database containing the supporting data as well as metadata for its processing.

The process that has been followed for a proper metadata support has been deep analysis of each process with special emphasis on input data, output data, constraints and requirements for its execution.This is the key basis of the approach. In particular, figures 4 and 5 depicts the part of the webhouse dedicated to data related to

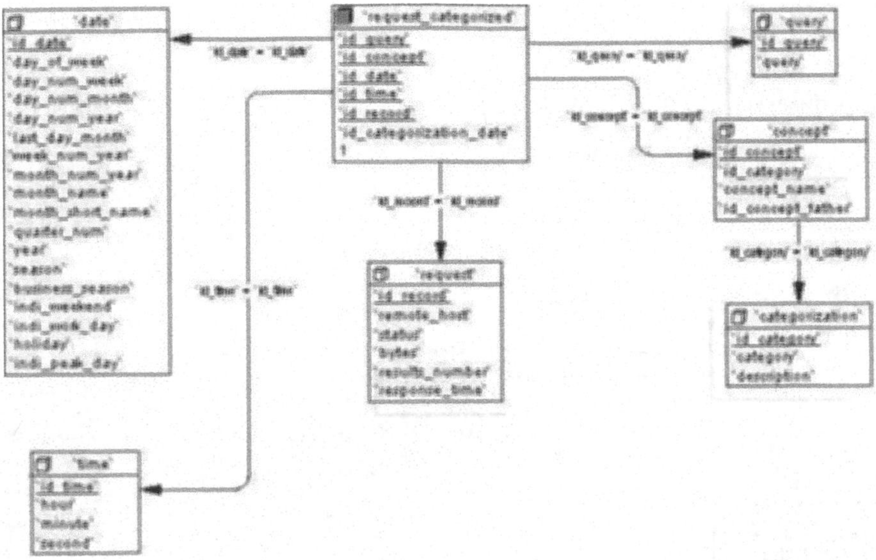
'date'
'id_date'
'day_of_week'
'day_num_week'
'day_num_month'
'day_num_year'
'last_day_month'
'week_num_year'
'month_num_year'
'month_name'
'month_short_name'
'quarter_num'
'year'
'season'
'business_season'
'indi_weekend'
'indi_work_day'
'holiday'
'indi_peak_day'
'request_categorized'
'id_query'
'id_concept'
'id_date'
'id_time'
'id_record'
'id_categorization_date'
'query'
'id_query'
'query'
'concept'
'id_concept'
'id_category'
'concept_name'
'id_concept_father'
'request'
'id_record'
'remote_host'
'status'
'bytes'
'results_number'
'response_time'
'categorization'
'id_category'
'category'
'description'
'time'
'id_time'
'hour'
'minute'
'second'

Fig. 4 WebHouse: request submodel

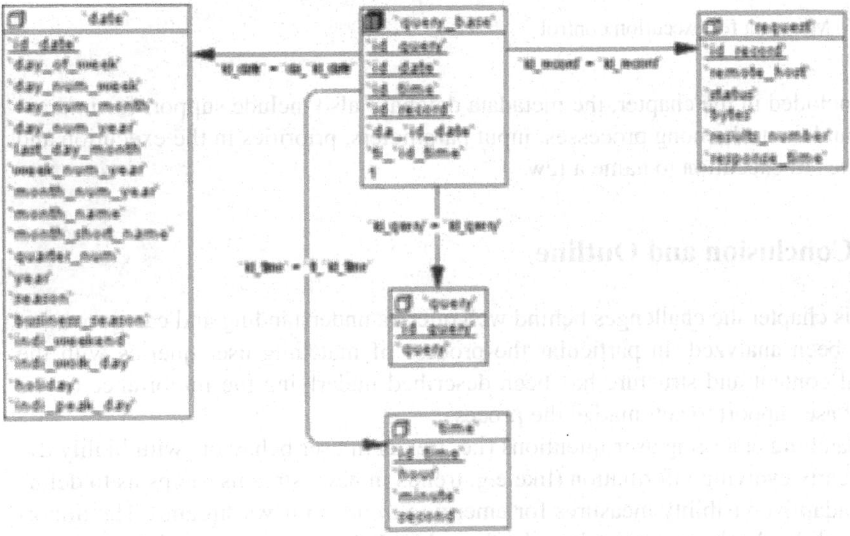
'date'
'id_date'
'day_of_week'
'day_num_week'
'day_num_month'
'day_num_year'
'last_day_month'
'week_num_year'
'month_num_year'
'month_name'
'month_short_name'
'quarter_num'
'year'
'season'
'business_season'
'indi_weekend'
'indi_work_day'
'holiday'
'indi_peak_day'
'query_base'
'id_query'
'id_date'
'id_time'
'id_record'
'request'
'id_record'
'remote_host'
'status'
'bytes'
'results_number'
'response_time'
'query'
'id_query'
'query'
'time'
'id_time'
'hour'
'minute'
'second'

Fig. 5 WebHouse: query submodel

query labeling and visibility calculation for a particular case of a portal dedicated to news. On the other hand, figure 6 depicts the part dealing with information to control each execution of each process of the preprocessing processes. Although

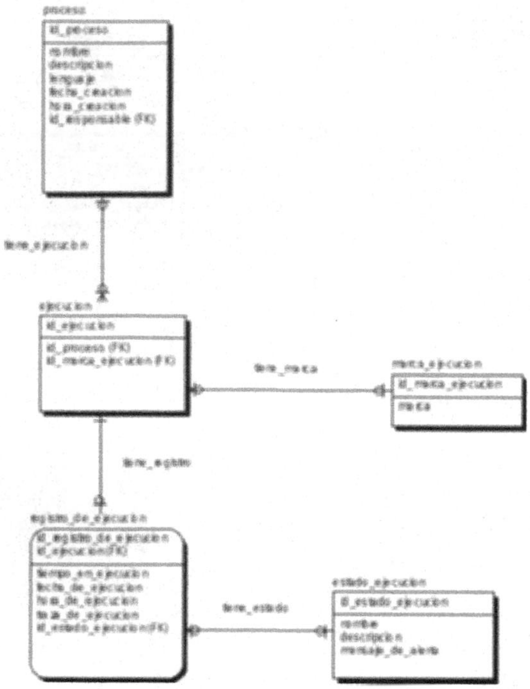

Fig. 6 Metadata for execution control

not included in the chapter, the metadata database also include support for interaction and control among processes, input parameters, priorities in the execution and policies of execution to name a few.

4 Conclusion and Outline

In this chapter the challenges behind web queries understanding and categorization have been analyzed. In particular the process of matching user queries with the portal content and structure has been described underlying the importance of the data base support to automatize the process.

Matching emerging user intentions (i.e. trends in user behavior) with highly dynamically evolving information (like e.g. trends in news streams) helps us to define user-adaptive visibility measures for emerging trends in news streams. The importance of the database support has also been highlighted and one possible design of underlying database have been described in the chapter. The proposed approach is only the first starting point on how to optimize the matching of user query terms to the emerging topics extracted from the structure of the portal where they are located. Future work on this direction will make it possible to find more effective ways of modeling and optimizing the synchronization process of trends detection and user behavior.

References

1. Annand, S.: Putting the user in context, 2006. In: ECML PKDD 2006 Workshop on Ubiquitous Knowledge Discovery for users (UKDU 2006), Berlin (2006)
2. Beeferman, D., Berger, A.: Agglomerative clustering of a search engine query log. In: KDD 2000: Proceedings of the sixth ACM SIGKDD international conference on Knowledge discovery and data mining, pp. 407–416. ACM, New York (2000)
3. Beitzel, S.M.: On understanding and classifying web queries. PhD Thesis, Illinois Institute of Technology (2006)
4. Beitzel, S.M., Jensen, E.C., Chowdhury, A., Frieder, O., Grossman, D.: Temporal analysis of a very large topically categorized web query log. J. Am. Soc. Inf. Sci. Technol. 58(2), 166–178 (2007)
5. Beitzel, S.M., Jensen, E.C., Frieder, O., Grossman, D., Lewis, D.D., Chowdhury, A., Kolcz, A.: Automatic web query classification using labeled and unlabeled training data. In: SIGIR 2005: Proceedings of the 28th annual international ACM SIGIR conference on Research and development in information retrieval, pp. 581–582. ACM Press, New York (2005)
6. Beitzel, S.M., Jensen, E.C., Lewis, D.D., Chowdhury, A., Frieder, O.: Automatic classification of web queries using very large unlabeled query logs. ACM Trans. Inf. Syst. 25(2), 9 (2007)
7. Bollegala, D., Matsuo, Y., Ishizuka, M.: Measuring semantic similarity between words using web search engines. In: WWW 2007: Proceedings of the 16th international conference on World Wide Web, pp. 757–766. ACM Press, New York (2007)
8. Broder, A.: A taxonomy of web search. SIGIR Forum 36(2), 3–10 (2002)
9. Chung, S., McLeod, D.: Dynamic topic mining from news stream data. In: CoopIS/DOA/ODBASE, pp. 653–670 (2003)
10. Dumais, S.T., Chen, H.: Hierarchical classification of Web content. In: Belkin, N.J., Ingwersen, P., Leong, M.-K. (eds.) Proc. of SIGIR-2000, 23rd ACM International Conference on Research and Development in Information Retrieval, Athens, GR, pp. 256–263. ACM Press, New York (2000)
11. Eibe, S., Valencia, M., Menasalvas, E., Segovia, J., Sousa, P.: Towards user context enhance search engine logs mining. In: Proceedings of the AWIC 2007 (2007)
12. Gravano, L., Hatzivassiloglou, V., Lichtenstein, R.: Categorizing web queries according to geographical locality. In: 12th ACM Conference on Information and Knowledge Management (CIKM 2003), November 3-8, pp. 325–333. ACM Press, New York (2003)
13. Jansen, B.J., Booth, D.L., Spink, A.: Determining the user intent of web search engine queries. In: WWW 2007: Proceedings of the 16th international conference on World Wide Web, pp. 1149–1150. ACM, New York (2007)
14. Jansen, B.J., Spink, A.: How are we searching the world wide web? a comparison of nine search engine transaction logs. Inf. Process. Manage. 42(1), 248–263 (2006)
15. Jansen, B.J., Spink, A., Saracevic, T.: Real life, real users, and real needs: a study and analysis of user queries on the web. Inf. Process. Manage. 36(2), 207–227 (2000)
16. Jones, R., Diaz, F.: Temporal profiles of queries. ACM Trans. Inf. Syst. 25(3), 14 (2007)
17. Joshi, H., Ito, S., Kanala, S., Hebbar, S., Bayrak, C.: Concept set extraction with user session context. In: ACM-SE 45: Proceedings of the 45th annual southeast regional conference, pp. 455–460. ACM, New York (2007)
18. Kang, I., Kim, G.: Query type classification for web document retrieval (2003)
19. Kawai, Y., Kumamoto, T., Tanaka, K.: User preference modeling based on interest and impressions for news portal site systems. In: Bressan, S., Küng, J., Wagner, R. (eds.) DEXA 2006. LNCS, vol. 4080, pp. 549–559. Springer, Heidelberg (2006)

20. Kontostathis, A., Galitsky, L., Pottenger, W.M., Roy, S., Phelps, D.J.: A Survey of Emerging Trend Detection in Textual Data Mining. Springer, Heidelberg (2003)
21. Kukulenz, D., Ntoulas, A.: Answering bounded continuous search queries in the world wide web. In: WWW 2007: Proceedings of the 16th international conference on World Wide Web, pp. 551–560. ACM, New York (2007)
22. Kules, B., Kustanowitz, J., Shneiderman, B.: Categorizing web search results into meaningful and stable categories using fast-feature techniques. In: JCDL, pp. 210–219 (2006)
23. Lee, U., Liu, Z., Cho, J.: Automatic identification of user goals in web search. In: WWW 2005: Proceedings of the 14th international conference on World Wide Web, pp. 391–400. ACM, New York (2005)
24. Li, Y.: Mining ontology for automatically acquiring web user information needs. IEEE Transactions on Knowledge and Data Engineering 18(4), 554–568 (2006) (Senior Member-Ning Zhong)
25. Rose, D.E., Levinson, D.: Understanding user goals in web search. In: WWW 2004: Proceedings of the 13th international conference on World Wide Web, pp. 13–19. ACM, New York (2004)
26. Shen, D., Pan, R., Sun, J.-T., Pan, J.J., Wu, K., Yin, J., Yang, Q.: Query enrichment for web-query classification. ACM Trans. Inf. Syst. 24(3), 320–352 (2006)
27. Shen, D., Sun, J.-T., Yang, Q., Chen, Z.: Building bridges for web query classification. In: SIGIR 2006: Proceedings of the 29th annual international ACM SIGIR conference on Research and development in information retrieval, pp. 131–138. ACM, New York (2006)
28. Sieg, A., Mobasher, B., Burke, R.D.: Representing context in web search with ontological user profiles. In: Kokinov, B., Richardson, D.C., Roth-Berghofer, T.R., Vieu, L. (eds.) CONTEXT 2007. LNCS, vol. 4635, pp. 439–452. Springer, Heidelberg (2007)
29. Song, R., Luo, Z., Wen, J.-R., Yu, Y., Hon, H.-W.: Identifying ambiguous queries in web search. In: WWW 2007: Proceedings of the 16th international conference on World Wide Web, pp. 1169–1170. ACM, New York (2007)
30. Spink, A., Jansen, B.J., Blakely, C., Koshman, S.: Overlap among major web search engines. In: ITNG 2006: Proceedings of the Third International Conference on Information Technology: New Generations (ITNG 2006), Washington, DC, USA, pp. 370–374. IEEE Computer Society, Los Alamitos (2006)
31. Wen, J.-R., Nie, J.-Y., Zhang, H.: Clustering user queries of a search engine. In: WWW, pp. 162–168 (2001)

Applications of Fuzzy and Rough Set Theory in Data Mining

Dan Li and Jitender S. Deogun

The explosion of very large databases has created extraordinary opportunities for monitoring, analyzing and predicting global economical, geographical, demographic, medical, political, and other processes in the world. Statistical analysis and data mining techniques have emerged for these purposes. Data mining is the process of discovering previously unknown but potentially useful patterns, rules, or associations from huge quantity of data. Data mining can be performed on different data repositories such as relational databases, data warehouses, transactional databases, sequence databases, spatial databases, spatio-temporal databases, and text databases, etc. Typically, data mining functionalities can be classified into two categories: descriptive and predictive. Descriptive mining tasks aim at characterizing the general properties of the data in the databases, while predictive mining tasks perform inherence on the current data in order to make prediction in future.

When our knowledge extraction task involves numerical data with continuous attributes, each attribute is usually discretized into several intervals. The discretization into intervals is used in many data mining techniques such as decision trees. In some situations, human knowledge exactly corresponds to such discretization of continuous attributes. However, in other situations, the discretization into intervals is not appropriate for describing human knowledge. For example, we may have the following knowledge in intrusion detection: "When the duration of a connection is very short, this connection is thought as a normal activity". We cannot appropriately represent this knowledge using the discretization of the domain of connection

Dan Li
Department of Computer Science, Northern Arizona University,
Flagstaff AZ 86011-5600

Jitender S. Deogun
Department of Computer Science and Engineering, University of Nebraska-Lincoln,
Lincoln NE 68588-0115

D. Zakrzewska et al. (Eds.): Meth. and Support. Tech. for Data Analys., SCI 225, pp. 71–113.
springerlink.com

duration into intervals. This is because the term “very short” cannot be appropriately represented by an interval. Given such limitations in current data mining techniques, there is a trend of integrating soft computing techniques into data mining methodologies to handle imprecision, uncertainty, and approximation in data. Soft computing differs from conventional (hard) computing in that, unlike hard computing, it is more tolerant of imprecision, uncertainty, partial truth, and approximation. The key areas of soft computing include neural networks, fuzzy logic, evolutionary computation, and rough set theory. This chapter explores three real-world applications which show the improvement to the existing data mining solutions by integrating two soft computing paradigms, fuzzy logic and rough set theory.

This chapter is organized into four sections. Section 1 introduces basic concepts related to data mining. It discusses the needs of data mining and explores the functionalities of data mining. Section 2 shows the concept of soft computing and soft computing solutions based on fuzzy set theory and rough set theory. Section 3 addresses the needs of integrating soft computing into data mining and explores the applications of fuzzy logic and rough set theory in two data mining tasks, i.e., classification and clustering. Finally, Section 4 summarizes the entire chapter.

1 Data Mining

It is estimated that the amount of information in the world doubles every 20 months [17]; that is, many scientific, government and corporate information systems are being overwhelmed by a flood of data that is generated and stored routinely, which grows into large databases amounting to giga (and even tera) bytes of data. These databases contain potential gold mine of valuable information, but it is beyond human ability to analyze such massive amounts of data and elicit meaningful patterns. Given certain data analysis goal, it has been a common practice to either design a database application on on-line data or use a statistical (or an analytical) package on off-line data along with a domain expert to interpret the results. Even if one does not count the problems related with the use of standard statistical packages (such as its limited power for knowledge discovery, the needs for trained statisticians and domain experts to apply statistical methods and to refine/interpret results, etc.), one is required to state the goal (i.e., what kind of information one wishes to extract from data) and gather relevant data to arrive at that goal. Consequently, there is still strong possibility that some significant and meaningful patterns in the database, waiting to be discovered, are missed.

As often argued in the literature it is desirable to pursue a more general goal, which is to extract implicit, previously unknown, hidden, and potentially useful information from raw data in an automatic fashion, rather than developing individual applications for each user need. Unfortunately, the database technology of today offers little functionality to explore data in such a fashion. At the same time knowledge discovery techniques for intelligent data analysis are not yet mature for large data sets [49]. Furthermore, the fact that data has been organized and collected around the needs of organizational activities may pose a real difficulty in locating the relevant data for knowledge discovery techniques from diverse sources. The data mining problem is defined to emphasize the challenges of searching for knowledge in large databases and to motivate researchers and application developers for meeting that challenge. It comes from the idea that large databases can be viewed as data mines containing valuable information that can be discovered by efficient knowledge discovery techniques.

1.1 Data Mining and Related Fields

Data mining is a promising interdisciplinary area of research shared by several fields such as database systems, machine learning, intelligent information systems, statistics, data warehousing, and knowledge acquisition in expert systems [21]. It may be noted that data mining is a distinct discipline and its objectives are different from the goals and emphases of the individual fields. Data mining may, however, heavily use theories and developments of these fields. In the following, we present basic differences (and/or similarities) between data mining and various allied research areas.

We have seen an increasing use of techniques in data mining that draw upon or are based on statistics in the last few years; namely, in feature selection [15], classification of objects based on descriptions [13], discretization of continuous values [22], prediction of missing values [19], etc. However, the statistical methods are limited either because of strong statistical assumptions, such as adherence to a particular probability distribution model, or due to its inability to recognize and generalize relationships, such as the set inclusion, that capture structural aspects of a data set, as a result of being entirely confined to arithmetic manipulations of probability measures [66].

In the earlier work on machine learning, a number of theoretical and foundational issues of interest to data mining have been investigated, e.g., learning from examples, formation of concepts from instances, discovering regular patterns, noisy and incomplete data manipulation, and uncertainty management. Data mining simply combines all aspects of knowledge discovery in the context of ultra large data. More specifically, data mining is the process of deriving rules, where a database takes on the role of training data set. In other words, a data mining application distinguishes itself from machine learning problem, in the sense that available techniques must be extended to be applicable to uncontrolled, real-world data. That is, one does not have the luxury of specifying the data requirements from the perspective of knowledge discovery goals before collecting the data.

1.2 Data Mining Functionalities

Data mining functionalities are used to specify the kind of patterns to be found in data mining tasks. Typically, data mining tasks can be classified into two categories: *descriptive* and *predictive*. Descriptive mining tasks aim at characterizing the general properties of the data in the databases, while predictive mining tasks perform inherence on the current data in order to make prediction in future [28]. In some cases, users may have no idea regarding what kinds of patterns in their data may be of interest, and hence may like to search for several different kinds of patterns in parallel. Thus, it is important to have a data mining system that can mine multiple kinds of patterns to accommodate different user expectations or applications.

In the rest of this section, we explore the main functionalities of data mining and the kinds of patterns a user can discover.

1.2.1 Association Rule Mining

Association rule mining is one of the descriptive tasks in data mining. Since its introduction in 1993 [3], it has received a great deal of attention. Today, association rule mining is still one of the popular pattern discovery methods in Knowledge Discovery in Databases (KDD). Formally, given a set of transactions, each of which contains some number of items from a given data set, *association rule mining* is to produce dependency rules which will describe the presence of items based on the presence of other items.

Normally, the rules have the form of $X \Rightarrow Y$, where X is the rule antecedent, Y is the rule consequent, and $X \cap Y = \emptyset$. *Support* and *confidence* are two widely used metrics in measuring the interestingness of association rules. The support of a rule $X \Rightarrow Y$ is denoted by $sup(X \Rightarrow Y)$. It indicates the percentage of items in the data set that contain both X and Y. We can see that support is simply a measure of its statistical significance [29]. The confidence of a rule $X \Rightarrow Y$ is denoted by $conf(X \Rightarrow Y)$, which indicates the possibility that a transaction contains Y given that it contains X. It is defined as $conf(X \Rightarrow Y) = sup(X \Rightarrow Y)/sup(X)$.

Association rule mining has its wide applications in marketing and sales promotion. For instance, suppose we have such an association rule *bagel* $\Rightarrow$ *potato_chip*. Since potato_chip is the consequent of the rule, this information can be used to determine what should be done to boost its sales. On the other hand, since bagel is the rule antecedent, this information can be used to determine which products would be affected if the store discontinues selling bagels. To find all the interesting association rules, each rule must satisfy both a *minimum support threshold* and a *minimum confidence threshold*, which are typically specified by users through statistical analysis.

1.2.2 Classification for Prediction

Classification is the process of finding a model (or function) that describes and distinguishes data classes or concepts, for the purpose of being able to use the model to predict the class of objects whose class label is unknown. The derived (or classification) model is based on the analysis of a set of training data (i.e., data objects whose class label is known) [28]. In other words, the derived classification model is a predictive model, i.e., a mapping from observations about an item to conclusions about its target value. Usually, a test data set (i.e., data objects whose class label is unknown) is used to determine the accuracy of the derived model.

The classification model may be represented in various forms, such as IF-THEN rules, decision trees, mathematical formulae, or neural networks. A *decision tree* (also called classification tree or regress tree) is a flow-chart-like tree structure, where leaves represent classifications and branches represent conjunctions of features (or attributes) that lead to those classifications. Decision trees can easily be converted to classification rules. A *neural network*, when used for classification, is typically a collection of neuron-like processing units with weighted connections between the units. There are many other methods for building classification models, such as Naïve Bayesian classification, Support Vector Machines, and K-Nearest Neighbor classification. Some of these methods will be introduced in later sections.

While classification predicts categorical (discrete, unordered) class labels, *regression analysis* is a statistical methodology that is most often used for numeric prediction. It predicts a value of a given continuous valued variable based on the values of other variables, assuming a linear or nonlinear model of dependency. Many problems can be solved by linear regression, and even more can be tackled by applying transformation to the variables so that a nonlinear problem can be converted to a linear one [28].

1.2.3 Cluster Analysis

Given a set of data points, each having a set of attributes, and a similarity measure among them, *cluster analysis* is to find clusters such that data points in one cluster are more similar to one another and data points in separate clusters are less similar to one another. Unlike classification, which analyze class-labeled data objects, clustering analyzes data objects without consulting a known class label. In general, the class labels are not available in the training data simply because they are not known to begin with. Clustering can be used to generate such labels. The overall goal of clustering is to divide data objects into several clusters or groups such that the intra-cluster similarity is maximized and the inter-cluster similarity is minimized.

Cluster analysis has been widely used in numerous applications, including market analysis and research, pattern recognition, data analysis, and image processing. Clustering can also be used for outlier detection, where outliers (values that are "far way" from any cluster) may be more interesting than common cases. Applications of outlier detection include credit card fraudulence detection, network intrusion detection (which will be introduced in later section), and the monitoring of other criminal activities in E-commerce.

Clustering is a challenging field of research in which its potential applications pose their own special requirements. For example, scalability is an important research topic in cluster analysis. Many clustering algorithms work well on small-scale data sets containing fewer than several hundred data objects; however, a large-scale data set may contain millions of objects. Another issue in clustering is the ability to deal with noisy data. Most real-world databases contain outliers or missing, unknown, or erroneous data. Some clustering algorithms are sensitive to such data and may lead to clusters of poor quality. In addition, choosing an appropriate similarity measure is the key of cluster analysis. For instance, Euclidean Distance is one of the most widely used similarity measures which are designed to cluster interval-based (numerical data). However, some applications may require clustering other type of data, such as binary, categorical (nominal), ordinal data, or mixtures of these data types. In these cases, there is a need to develop a similarity metric which is able to deal with different types of attributes.

2 Fuzzy Set Theory and Rough Set Thoery

The guiding principle of soft computing is to exploit the tolerance for imprecision, uncertainty, partial truth, and approximation to achieve tractability, robustness and low solution cost [64]. The key areas of soft computing include neural networks, fuzzy set theory, evolutionary computation, and rough set theory. In the rest of this section, we introduce two soft computing paradigms, fuzzy set theory and rough set theory.

2.1 An Introduction to Fuzzy Set Theory

Fuzzy logic has been recognized as a convenient tool for handling continuous attributes in a human understandable manner. The essential characteristics of fuzzy logic as founded by Lotfi A. Zadeh [64] are as follows:

- In fuzzy logic, exact reasoning is viewed as a limiting case of approximate reasoning.
- In fuzzy logic, everything is a matter of degree.
- Any logical system can be fuzzified.
- In fuzzy logic, knowledge is interpreted as a collection of elastic or, equivalently, fuzzy constraint on a collection of variables.
- Inference is viewed as a process of propagation of elastic constraints.

Fuzzy set theory is a set theory based on fuzzy logic. It is an extension of the classical notion of set. In classical set theory, the membership of elements in a set is assessed in binary terms according to a bivalent condition, i.e., an element either belongs or does not belong to the set. By contrast, fuzzy set theory permits the gradual assessment of the membership of elements in a set. The fuzzy sets of attributes interpret the value of an attribute as a membership degree (between 0 and 1) that determines to what extent the example is described by the attribute. In other words, an object can be entirely in the set (if membership degree = 1), entirely not in the set (if membership degree = 0), or partially in the set (if $0 <$ membership degree < 1). The membership degree of an object defines a function in which the universe of the set of values that the object can take is the domain, and the interval [0, 1] is the range. This function is called *membership function*. Figure 1 shows the commonly used membership function, the triangular membership function. In this figure, the object x has 0.73 degree of membership to the fuzzy set *low*.

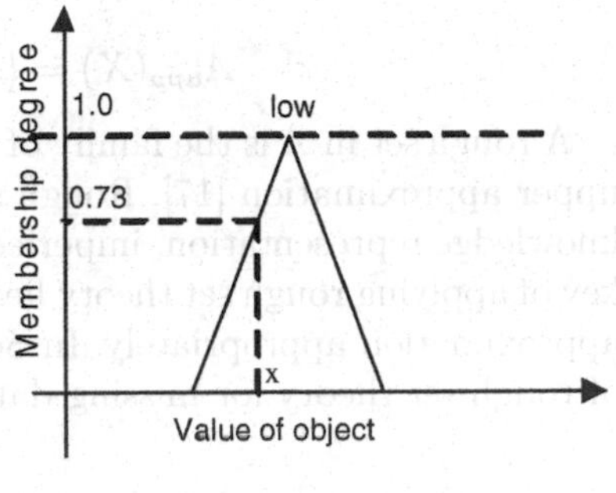
Membership degree
1.0
low
0.73
x
Value of object

Fig. 1 Triangular membership function for a fuzzy set

2.2 An Introduction to Rough Set Theory

Even though it has been more than two decades since the introduction of the rough set theory by Zdzislaw Pawlak [52], there is still a continued need for further development of rough functions and for extending rough set model to new applications. We believe that the investigation of the rough set methodology for data mining is a challenging research area with promise of high payoffs in many business and scientific domains. Additionally, such investigations will lead to the integration of the rough set theory with other knowledge discovery methodologies, under the umbrella of data mining applications.

The rough set philosophy is founded on the assumption that, with every object of the universe of discourse, we associate some information (data, knowledge). Objects characterized by the same information are indiscernible in view of the available information about them. The indiscernibility relation generated in this way is the mathematical basis for the rough set theory. Any set of all indiscernible objects is called an elementary set and forms a basic granule of knowledge about the universe. Any set of objects being a union of some elementary sets is referred to as crisp (precise); otherwise, a set is rough (imprecise, vague). Consequently, each rough set has boundary-line cases, i.e., objects which cannot be classified with certainty as members of the set or of its complement. Therefore, a rough set can be replaced by a pair of crisp sets, called the lower and the upper approximation. The lower approximation consists of all objects which surely belong to the set and the upper approximation contains objects which possibly belong to the set [59].

Formally, let the pair $A = (U, R)$ be an *approximation space*, where U is a finite nonempty set of objects, called *universe*, and R is a set of equivalence classes on U, called *indiscernibility relation.* A s*definable set* in A is obtained by applying a finite number of union operations on R. For $x \in U$, let $[x]_R$ denote the equivalence class of R, containing x. For each $X \subseteq U$, X is characterized in A by a pair of sets — its *lower* and *upper approximations* in A, defined respectively as:

$$A_{low}(X) = \{x \in U | [x]_R \subseteq X\}$$

$$A_{upp}(X) = \{x \in U | [x]_R \cap X \neq \emptyset\}$$

A rough set in A is the family of all subsets of U having the same lower and upper approximation [17]. Rough set theory has been used for data mining, knowledge representation, imperfect data handling, and data reduction. The key of applying rough set theory lies in how to identify the lower and the upper approximation appropriately. In Section 3.3, we will explore the application of rough set theory for missing data manipulation.

3 Integrating Fuzzy and Rough Set Theory into Data Mining

When our knowledge extraction task involves numerical data with continuous attributes, each attribute is usually discretized into several intervals [33]. The discretization into intervals is used in many learning techniques such as decision trees. In some situations, human knowledge exactly corresponds to such discretization of continuous attributes. For example, the domain of connection times from the same IP address is divided into two intervals by the threshold 10,000 in the following pattern: "If a host receives more than 10,000 connection requests from the same IP address within one second, this IP address is suspicious of intrusion attack". However, in other situations, the discretization into intervals is not appropriate for describing human knowledge. For example, we may have the following knowledge in intrusion detection: "When the duration of a connection is very short, this connection is thought as a normal activity". We cannot appropriately represent this knowledge using the discretization of the domain of connection duration into intervals. This is because the term "very short" cannot be appropriately represented by an interval.

If we choose a quantitative measurement, a range value or an interval to denote a normal value, any values falling outside the interval will be considered outliers to the same degree regardless of their variations in distances to the interval. The same situation applies to values inside the interval, i.e., all will be viewed as normal to the same degree. This problem has been recognized as "sharp boundary" problem [40]. Many outlier detection systems assume the existence of sharp boundary between normal and anomalous behavior. This assumption, consequently, causes an abrupt separation between normalcy and anomaly. However, the normalcy is a vague concept. A natural way to characterize the normal is by defining a degree of normalcy. Therefore, a better characterization of the boundary between normal and abnormal is needed to increase the accuracy in the detection of outliers. As we know, the soft computing paradigm is to exploit the tolerance for imprecision, uncertainty and partial truth to achieve tractability, robustness and low solution cost. Thus, there is a need of integrating soft computing techniques into the process of data mining to address the above uncertainty issues.

Fuzzy logic and rough set theory are two key areas of soft computing. Recent research has shown that fuzzy and rough set theory has been applied to every aspect of data mining, including data cleaning and preprocessing, pattern discovery, and pattern interpretation and evaluation. Feature selection, also known as variable selection, feature reduction, attribute selection, or variable subset selection, is a commonly used data preprocessing technique in machine learning. It selects a subset of relevant features for building robust learning models. Recently, there is a trend of integrating fuzzy and rough set theory into the process of feature selection to eliminate redundant features from the data sets. In [44], the redundant attributes are removed by identifying the closure of an equivalence class, called *congruence class*. This class of features reduces the size of the original data set while it still maintains the same amount of useful information as the original data set. In [65], a fuzzy approach is used to deal with incomplete or imprecise data sets. An equivalence class is then defined based on the concepts of rough set to further eliminate redundant or insignificant attributes from the data sets.

Fuzzy and rough-based can also be applied to the process of pattern discovery. This includes association rule mining, classification, and cluster analysis, as described in Section 1. In [32], the author shows that fuzzy associations can be interpreted in different ways and that the interpretation has a strong influence on their assessment and, hence, on the process of rule mining. The author uses multiple-valued implication operators in order to model fuzzy association rules and propose quality measures suitable for this type of rules. Furthermore, the authors introduce a semantic model of fuzzy association rules which considers association rules as a convex combination of simple association rules. This model provides a sound theoretical basis and gives an explicit meaning to fuzzy associations.

Both fuzzy set and rough set theory can be integrated into the process of cluster analysis. A fuzzy-based clustering approach is presented in [36]. Traditional clustering methods using Euclidean distance can only identify clusters in simple (hyper-)spheres. The paper [36] focuses on a modified fuzzy distance function based on dot production which allows the discovery of clusters in more advanced shapes, for instance, ellipsoids, lines or shells of circles, and ellipses. Paper [39] presents a new indiscernibility-based rough agglomerative hierarchical clustering algorithm for sequential data. In this approach, the indiscernibility relation has been extended to a tolerance relation with the transitivity property being relaxed. Initial clusters are formed using a similarity upper approximation. Subsequent clusters are formed using the concept of constrained-similarity upper approximation wherein a condition of relative similarity is used as a merging criterion. The rough clusters resulting from the proposed algorithm provide interpretations of different navigation orientations of users present in the sessions without having to fit each object into only one group. Such descriptions can help web miners to identify potential and meaningful groups of users.

In the rest of this section, we provide three real-world projects which show the integration of fuzzy logic and rough set theory into current data mining solutions. The first two projects apply fuzzy set theory to two classification methods, Naïve Bayesian classification [42] and K-Nearest Neighbor classification [43], respectively. The third project combines fuzzy set theory and rough set theory and applies its solution to k-means clustering method for missing data imputation[41].

3.1 APPLICATION 1: Fuzzy Bayesian Classification for Anomaly Detection

3.1.1 Anomaly Detection — Problem Description

Due to increasing incidents of cyber attacks and heightened concerns for cyber terrorism, implementing effective intrusion detection systems (IDSs) is an essential task for protecting cyber security. Intrusion detection is the process of monitoring and analyzing the events occurring in a computer system in order to detect signs of security problems [7]. Intrusion detection starts with instrumentation of a computer network for data collection. Pattern-based software "sensors" monitor the network traffic and raise "alarms" when the traffic matches one of the known malicious patterns. This technique is called *misuse detection* [34]. Thus, misuse detection has the ability to identify intrusions based on a known pattern for the malicious activity. The pattern may be a static bit string, for example, a specific virus bit string insertion. Alternatively, the pattern may describe a suspect set or sequence of activities. These known patterns are referred to as *signatures*. A key advantage of misuse detection techniques is their high degree of accuracy in detecting known attacks and their variations [18]. However, the disadvantage of signature-based detection methods is that they can only detect previously known intrusion types since these intrusions have a corresponding signature. The signature database has to be manually updated for each new type of attack that is discovered and without this update, systems are vulnerable to these attacks.

Using data mining techniques to build profiles for anomaly detection has been an active research area for network intrusion detection. The ADAM [9, 10] has been recognized as the most widely known and well-published project in this field. Traditionally, anomaly detection methods require training over clean data (normal data containing no anomalies) in order to build a model that detects anomalies. There are two inherent drawbacks of these systems. First, clean training data is not always easy to obtain. Second, training over imperfect (noisy) data may result in systems accepting intrusive behavior as normal. To address these weaknesses, the possibility of training anomaly detection systems over noisy data has been investigated recently [20, 54]. Methods for anomaly detection over noisy data do not assume that the data is labelled or somehow otherwise sorted according to classification. These systems usually make one key assumption about the training data, i.e., data instances having the same classification (type of attack or normal) should be close to each other in feature space under some reasonable metric. In other words, anomalous elements are assumed to be qualitatively different from the normal. This assumption degrades the performance of most anomaly detection system. To solve this problem, we develop a novel anomaly detection framework which integrates fuzzy logic to eliminate sharp boundary between normal and anomalous behavior.

3.1.2 Fuzzy Bayesian Classification

There are two main reasons to introduce fuzzy logic for intrusion detection. First, many quantitative features are involved in intrusion detection and can potentially be viewed as fuzzy variables. Second, security itself includes fuzziness [48]. In the rest of this section, we present how to apply fuzzy logic to Bayesian classification for anomaly detection.

Let C denote a class attribute with a finite domain $dom(C)$ of m classes, and $V_1, ..., V_n$ a number of attributes with finite domains $dom(V_1)$, ..., $dom(V_n)$. An *instance* i is described by its attribute values $v_1^i \in dom(V_1)$, ..., $v_n^i \in dom(V_n)$. Naïve Bayesian classifiers implement a probabilistic idea of classification — calculate the class of a new instance i by estimating for each class from $dom(C)$ the probability that the instance is in this class, and select the most probable class as the prediction of i. Formally, for all $c \in dom(C)$ they estimate the probability

$$P(C = c|V_1 = v_1^i, V_2 = v_2^i, ..., V_n = v_n^i) \tag{1}$$

that an instance i with the given attribute values has the class c. To simplify, we use $P(c|v_1^i, v_2^i, ..., v_n^i)$ to substitute the expression in (1).

The basic idea of Naïve Bayesian classification is to apply the Bayes theorem

$$P(Y|X) = \frac{P(X|Y)P(Y)}{P(X)}. \tag{2}$$

Since fuzzy logic has been recognized as a convenient tool for handling continuous attributes, we want to apply fuzzy logic to Naïve Bayesian classification for anomaly detection. In the fuzzy case, an instance i does not have exactly one value $v_j^i \in dom(V_j)$ for each attribute V_j, but has each value $v_j \in dom(V_j)$ to a degree $\mu_{v_j}^i \in [0,1]$, where the degree $\mu_{v_j}^i$ is determined by the membership function.

Before we go further, we want to normalize each numerical attribute in the data set so that the membership function for all the numerical attributes can be defined in the same way. In the normalization algorithm each numerical value in the data set is normalized between 0 and 1 according to the following equation:

$$x = \frac{x - MIN}{MAX - MIN},$$

where x is a numerical value, MIN is the minimum value for the attribute that x belongs to, and MAX is the maximum value for the attribute that x belongs to. For each numerical attribute, the triangular fuzzy membership function using five basic fuzzy sets {Low(L), Medium-Low(ML), Medium(M), Medium-High(MH), High(H)} is shown in Figure 2.

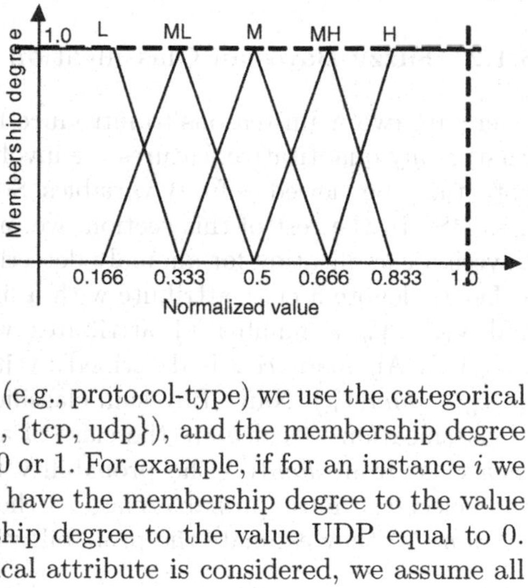
1.0
L
ML
M
MH
H
Membership degree
0.166
0.333
0.5
0.666
0.833
1.0
Normalized value

Fig. 2 Fuzzy function for numerical attributes

For any non-numerical attribute (e.g., protocol-type) we use the categorical values to construct a crisp set (e.g., {tcp, udp}), and the membership degree for each categorical value is either 0 or 1. For example, if for an instance i we have protocol-type = tcp, then we have the membership degree to the value TCP equal to 1 and the membership degree to the value UDP equal to 0. No matter a numerical or categorical attribute is considered, we assume all membership degrees are normalized for each instance i:

$$\sum_{v_j \in dom(V_j)} \mu_{v_j}^i = 1.$$

Given an instance i, we use $P(c|i)$ to denote the possibility that instance i belongs to class c, and use $P(v_j|i)$ to denote the possibility that instance i has an attribute v_j, i.e., $P(v_j|i) = \mu_{v_j}^i$. First, we split overall fuzzy cases for the actual attribute values, then we have

$$P(c|i) = \sum_{v_1 \in dom(V_1) \ldots v_n \in dom(V_n)} P(c|v_1 \ldots . v_n) P(v_1 \ldots v_n|i). \quad (3)$$

Since we assume that the attribute values of instance i are independent, the right-hand side of Equation (3) reduces to:

$$\begin{aligned} P(c|i) &= \sum_{v_1 \in dom(V_1) \ldots v_n \in dom(V_n)} P(c|v_1 \ldots . v_n) P(v_1|i) \ldots P(v_n|i) \\ &= \sum_{v_1 \in dom(V_1) \ldots v_n \in dom(V_n)} P(c|v_1 \ldots . v_n) \mu^i_{v_1} \ldots \mu^i_{v_n}. \quad (4) \end{aligned}$$

Now we apply Bayesian theorem, as shown in Equation (2), to Equation (4) and we obtain:

$$P(c|i) = \sum_{v_1 \in dom(V_1) \ldots v_n \in dom(V_n)} \frac{P(v_1 \ldots v_n|c) P(c)}{P(v_1 \ldots v_n)} \mu^i_{v_1} \ldots \mu^i_{v_n}. \quad (5)$$

The "Naïve" assumption made in Bayesian classification is that given the class value, all attribute values are independent. Although the independence assumption is Naïve in that it is in general not met, Naïve Bayesian classifiers give quite good results in many cases, and are often a good way to perform classification [60]. To deal with the dependencies among attribute values, one can apply a more sophisticated classification approach, *Bayesian Networks* [23].

We apply the same naïve independence assumption and finally obtain the following equation:

$$\begin{aligned} P(c|i) &= \sum_{v_1 \in dom(V_1) \ldots v_n \in dom(V_n)} \frac{P(v_1|c) \ldots P(v_n|c) P(c)}{P(v_1) \ldots P(v_n)} \mu^i_{v_1} \ldots \mu^i_{v_n} \\ &= P(c) \left(\sum_{v_1 \in dom(V_1)} \frac{P(v_1|c)}{P(v_1)} \mu^i_{v_1} \right) \ldots \left(\sum_{v_n \in dom(V_n)} \frac{P(v_n|c)}{P(v_n)} \mu^i_{v_n} \right). \quad (6) \end{aligned}$$

For intrusion detection, when we use Equation (6) to predict a testing instance i, $P(c|i)$ should be calculated for all $c \in dom(C) = \{normal, DoS, R2L, U2R, Probing\}$ to find the maximum value, p_{max}. If we have $p_{max} < \theta$, where θ is a user pre-specified possibility threshold, we assume a new type of attack occurs. The value of θ can be determined by empirical testing. In this way, fuzzy Bayesian classifiers facilitate the process of anomaly detection.

3.1.3 Preliminary Experimental Results

Experiments are conducted using 1998 DARPA intrusion detection data [46]. For each TCP connection, 41 various quantitative and qualitative features are extracted. We first apply naïve Bayesian classification to build the anomaly detection system based on all 41 features. Leave-One-Out (LOO) feature selection method is then applied to identify important, secondary, and unimportant features in feature space [42]. Finally, important and secondary features are used in the fuzzy Bayesian classification to detect anomalous behaviors.

Among 41 TCP connection features, by LOO feature selection method, four features are important because the removal of these features degrades the performance of the system considering two evaluation metrics, *F-value* and *accuracy*. Seven features are considered unimportant because the removal these features improves the overall performance of the detection system. The rest of 30 features are considered as secondary features because the overall performance is not affected by these features.

Table 1 shows the performance of the system based on four metrics, i.e., precision, recall, F-value, and accuracy [42]. We randomly divide the original test data set into 200 subsets. Each subset contains about 10,000 connection records. The experiments are conducted on each subset of the test data and the system is evaluated based on the average performance. From Table 1, the naïve Bayesian classification issues the worst performance, while the fuzzy Bayesian with feature selection renders the best results. The value of recall in the naïve classification is low because lots of real attacks are not successfully detected. However, after removing unimportant features from training and test data sets, the value of recall increases dramatically. Thus, with fewer number of features in the feature space, the system becomes more efficient and at the same time more accurate. Since there are 34 numerical attributes in the original feature set, the application of fuzzy logic represents imprecise knowledge precisely and improves the overall accuracy of the anomaly detection system.

Table 1 Comparison of Naïve and Fuzzy Bayesian Classifications

	Precision	Recall	F-value	OA
naïve w/o feature selection	94.7%	60%	1.47	97.5%
naïve w feature selection	93.3%	93.3%	1.87	99.2%
fuzzy w feature selection	96.5%	93.3%	1.89	99.5%

3.2 APPLICATION 2: Fuzzy K-Nearest Neighbor Classification for Gene Function Prediction

3.2.1 Gene Function Prediction — Problem Description

The availability of genome sequences of a variety of organisms has changed the world of biology. Identifying genes with respect to the function of their proteins has become essential for understanding the genomic information. One possible solution for identifying gene function is to eliminate or inhibit expression of a gene and observe any alterations in the phenotype. However, it is impractical to identify all possible functions for all genes by this approach. Thus, many statistical and data mining methods have been used to solve this problem. The most popular solution for gene function identification is BLAST [5].It predicts the functional class of a gene based on similar sequences for which the function is known. An obvious limitation of BLAST is that it cannot predict the function of a gene when no homologous gene of known function exists [57].

Recently, some advanced statistical and machine learning techniques have been employed for functional gene prediction. These include nearest neighbor approach [11, 58], Neural Networks [12], Markov Models [31], support vector machines [61], and decision trees [53]. Among all of these methods, Neural Networks have been proved to be one of the most accurate classification model for gene function prediction [57]. However, a major disadvantage of neural networks lies in their knowledge representation. The black box nature of a network makes it difficult for humans to interpret. Comparing with Neural Networks, K-Nearest Neighbor (KNN) approach has recently gained more popularity for building classification and prediction models, given its simplicity, easy interpretability, and acceptable accuracy rate.

Nearest neighbor classifiers are based on learning by analogy, that is, by comparing a given test tuple with training tuples that are similar to it. Since nearest neighbor has the same nature as BLAST, that is, the prediction is made based on similarity search, nearest neighbor approach also suffers the drawback of BLAST. It is not possible to identify a class value which does not exist in training data. Another disadvantage of original nearest neighbor approach is that all the selected neighbors are assigned the same weight, even though the similarity between a test tuple and its neighbors is different. Since fuzzy logic is a convenient tool for handling uncertainty in data sets, we apply fuzzy set theory to the original KNN classifier. While retaining the simplicity and easy interpretability of KNN approach, the proposed fuzzy system can be used to identify unknown class values and it improves the system accuracy by assigning different weights to the neighbors of a test case. Thus, this fuzzy classifier overcomes the two major drawbacks of the original KNN approach.

3.2.2 K-Nearest Neighbor Classification

A K-Nearest Neighbor (KNN) classifier is a lazy instance-based learning algorithm by similarity search [28]. It compares a given test object with training objects that are closest to it. The training objects are described by n attributes or features. Each object represents a points in a n-dimensional feature space. When given an unknown test object, the KNN classifier searches the feature space for the K training objects that are closest to the unknown test object. These K training objects are so-called "K-nearest neighbors" of the test object.

The "closeness" is defined in terms of a distance metric. In this paper, we use generalized L_P norm distance [4] to measure the distance between a test object x_i and a training object x_j,

$$d(x_i, x_j) = \left(\sum_{m=1}^{n} |x_{i,m} - x_{j,m}|^p \right)^{1/p} . \tag{7}$$

L_1 is called Manhattan distance and L_2 is called Euclidean distance. Another distance metric we choose is the Cosine-based distance which is calculated from Cosine Similarity,

$$d(x_i, x_j) = e^{-Sim(x_i, x_j)}, \tag{8}$$

where

$$Sim(x_i, x_j) = \frac{\sum_{m=1}^{n} x_{i,m} * x_{j,m}}{\sqrt{\sum_{m=1}^{n} {x_{i,m}}^2 \sum_{m=1}^{n} {x_{j,m}}^2}}.$$

Typically, we normalize the values of each attribute before using distance equations. This helps prevent attributes with initially large ranges from outweighing attributes with initially smaller ranges. Min-max normalization is used to transform a value.

For KNN classification, the unknown test object is assigned the most common class among its K nearest neighbors. When $K = 1$, the test object is assigned the class of the training object that is closest to it in feature space.

From the description of the algorithm we can see that KNN classifier cannot successfully identify the class for a test object if this class does not exist in the training data set. In addition, since the class prediction is made purely based on the maximal vote from K nearest neighbors, each neighbor has been assigned the same weight towards the vote. Given this feature, KNN classifier becomes extremely sensitive to the value of K.

For example, Figure 3 shows the first five nearest neighbors of a test object. If $K = 2$, the test object will be assigned to class *c1* since the two closest neighbors are both from class *c1*. If $K = 3$, the test object will also be assigned to class *c1* since among the three closest neighbors, two training objects belong to class *c1* and one belongs to class *c2*. To simplify, we assume K is an odd number to avoid equal votes among K neighbors. Now consider the case when $K = 5$, from KNN algorithm, we can easily conclude that the test object will be assigned to class *c2* as *c2* has the maximal vote, i.e., 3. However, when we refer to Figure 3, we find that the vote between classes *c1* and *c2* only differs by one, and the two most closest objects to the test case

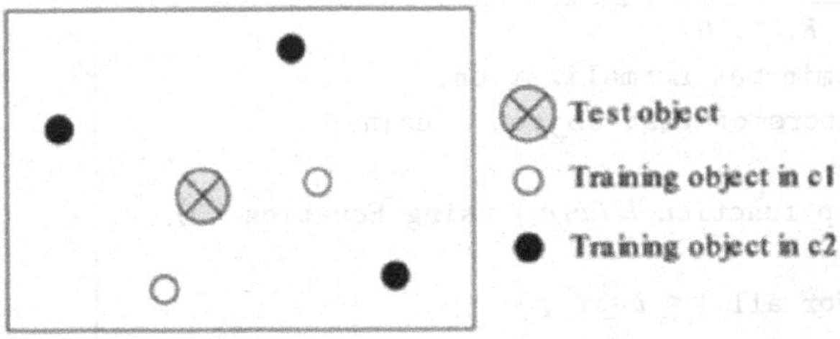
Test object
Training object in c1
Training object in c2

Fig. 3 K-Nearest Neighbor

are both from class *c1*. This indicates that the test object has more similar features with objects in class *c1*, even though class *c2* has more votes. To handle this uncertainty in the data set, we apply fuzzy set theory to original (crisp) KNN classification algorithm.

3.2.3 Integrating Fuzzy Set Theory into KNN

In fuzzy KNN classification, each test object x has a membership function which describes the degree that this data object belongs to certain class c_i. The membership function is defined as:

$$U(x, c_i) = \frac{\sum_{k=1}^{K} U(x_k, c_i) * d(x, x_k)^{-2/(m-1)}}{\sum_{k=1}^{K} d(x, x_k)^{-2/(m-1)}}, \tag{9}$$

where $m > 1$ is the fuzzifier which determines how heavily the distance is weighted when calculating each neighbor's contribution to the membership value, x_k is the kth nearest neighbors of object x, and K specifies the total number of nearest neighbors. $U(x_k, c_i)$ is 1 if the training object x_k belongs to class c_i, and 0 otherwise. Given the total number of classes is C in the training data set, we have $\sum_{i=1}^{C} U(x, c_i) = 1$ for any data object x.

From Equation (9) we can see that the fuzzy KNN approach overcomes the limitation of crisp KNN approach by considering the distance between the test case and each individual neighbor. This is based on the assumption that if two objects are close to each other in the feature space, they share the similar functionality. On the other hand, the fuzzy KNN classifier can be used to identify new class labels. If the maximal membership is less than a predefined threshold, θ, a new class label will be assigned to the test object and this object will also be added to the training data for future reference. This feature makes the fuzzy KNN method outperforms the most similarity-based classification methods. Figure 4 summarizes the fuzzy KNN algorithm.

```
Algorithm: Fuzzy-KNN(x, K, C, θ)
1) Normalize data using min-max normalization;
2) Find K nearest neighbors of test object x using
Equation (7) or (8);
3) Compute the membership function U(x,c_i) using Equation (9)
for each 1 ≤ i ≤ C;
4) Let c = max U(x,c_i) for all 1 ≤ i ≤ C;
5) If c ≥ θ, assign class c to test object x;
6) Otherwise, assign a new class label c' to test object x.
```

Fig. 4 Fuzzy K-Nearest Neighbor Classification

3.2.4 Experiments and Analysis

To predict the functional class of a gene, we need to collect gene expression profiles and functional classes of genes. We use the yeast gene expression profile published by Brown's group at Stanford University [1]. This data set includes 2467 yeast genes from 79 experiments. Since each experiment is performed on different subset of genes, there are some missing values among 2467 yeast genes. To simplify, we apply replace-with-mean approach to fill in the missing values. The functional annotations on yeast genes is obtained from Munich Information Center for Protein Sequence (MIPS) database [2]. This database defines a total of 249 gene classes in a hierarchical structure, i.e., some genes are assigned to multiple categories. Among 249 functional classes, nine largest functions are used for experiments [14]. Among 2467 yeast genes, 2331 genes can be categorized into these nine classes. Therefore, we use this subset of yeast genes to test the system. To summarize, our training data set consists of 2331 yeast genes of 79 attributes and they belong to nine gene classes.

We evaluate the performance of algorithms based on Leave-One-Out (LOO) cross validation resampling method, that is, one training sample is "left out" at a time for the test set. The beauty of cross validation is that each sample is used the same number of times for training and once for testing. This provides relatively low bias. The first performance metric we use is *accuracy*. The accuracy of a classifier on a given test set is the percentage of test set tuples that are correctly classified by the classifier, i.e.,

$$Accuracy = \frac{the\ number\ of\ true\ predictions}{the\ number\ of\ test\ records}.$$

The accuracy can be a biased performance metric for an unbalanced data set. For example, suppose that we have trained a classifier to classify test tuples as either "yes" or "no". An accuracy rate of, say, 95% makes the classifier seem quite accurate, but what if only 1-2% of the training tuples are actually "yes"? Clearly, the accuracy rate of 95% is still unacceptable in this case because the classifier might be correctly labeling only the "no" tuples. To address this concern, we use two more metrics to evaluate the algorithms, i.e., *macro-precision* and *macro-recall*.

The precision for a certain class c_i is defined as the percentage of correctly predicted records among all predicted records in class c_i. For macro-precision, the precision value for each class is calculated and the average value for all classes is calculated, i.e.,

$$Macro\text{-}precision = avg(\frac{true\ pos\ of\ c_i}{true\ pos + false\ pos\ of\ c_i}),$$

$$\text{for } i = 1, 2, ..., C.$$

The recall for a certain class c_i is defined as the percentage of correctly predicted records among all records in class c_i. For macro-recall, the recall value for each class is calculated and the average value for all classes is calculated, i.e.,

$$Macro\text{-}recall = avg(\frac{true\ pos\ of\ c_i}{true\ pos + false\ neg\ of\ c_i}),$$

$$\text{for } i = 1, 2, ..., C.$$

Figure 5 shows the system accuracy of crisp and fuzzy KNN classifiers as K the number of nearest neighbors varies. The upper picture is for Manhattan distance, while the lower picture shows the case when Euclidean distance is applied. We have two main observations. First, fuzzy KNN classifier provides higher accuracy than crisp KNN approach for all of the cases no matter how we set up the value of K. The difference becomes obvious when the value of K is greater than 15. Now, let's go back to Figure 3. This figure shows the scenario when a certain class c has the maximal vote among all the classes, but the training objects that belong to class c are not the most closest neighbors to the test object. By crisp KNN classifier, the test object will be assigned to class c without any other options. However, the fuzzy KNN classifier makes the prediction by considering both the number of closest neighbors and the distance between the test object and its neighbors. As the value of K increases, the training objects that are closer to the test object contribute more to the prediction result comparing with the training objects that are far away. The fuzzy KNN classifier, therefore, has a higher accuracy rate. The second observation from Figure 5 is that the best accuracy rate can be reached when K is between 15 and 25 and this is especially true when Manhattan distance metric is applied.

Figure 6 shows the system performance with regard to macro-precision and macro-recall while Manhattan distance and Euclidean distance are applied, respectively. Again, two observations. First, the fuzzy classifier does improve both the precision and the recall no matter how we set up input parameters, and it is consistent with Figure 5 that the improvement cannot be ignored while the value of K becomes larger. This, again, shows that integrating fuzzy set theory into original nearest neighbor classification algorithm has

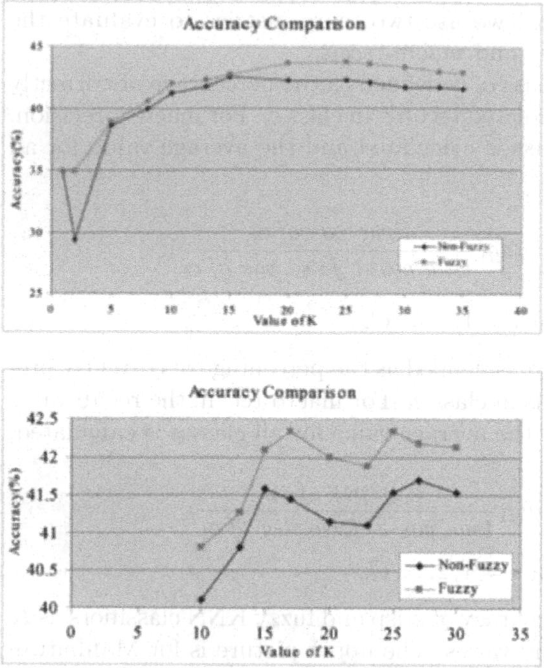
Accuracy Comparison
Accuracy(%)
Value of K
Non-Fuzzy
Fuzzy
Accuracy Comparison
Accuracy(%)
Value of K
Non-Fuzzy
Fuzzy

Fig. 5 Accuracy comparison when Manhattan distance and Euclidean distance are applied respectively

facilitated the process of handling uncertainty in the data sets. The second observation from Figure 6 is that, different from accuracy rate shown in Figure 5, the macro-precision and the macro-recall drop down dramatically as the value of K becomes greater than 25 if Manhattan distance is applied, and 20 if Euclidean distance is used. This exactly addresses the issue with accuracy measure. The reason for this result is that class distribution varies among nine classes presented in the training data set.

Table 2 summarizes the distribution of nine classes and their corresponding precision and recall for fuzzy classifier when $K = 20$. We notice that classes 2 and 5 have fewer number of records in the training data set, and consequently, these two classes generated poor precision and recall rate. Class 7 has the most number of instances in the data set and this class has a recall score of 0.9. What is the insight of these results? Class distribution cannot be ignored when nearest neighbor approach is employed for classification. It is reasonable that the instances belonging to higher ratio classes (for instance, class 7) have more chances to be selected as neighbors. This will result in lower precision and higher recall scores for such classes based on the definition of precision and recall. The result for Class 7 conforms to this conclusion. Then how

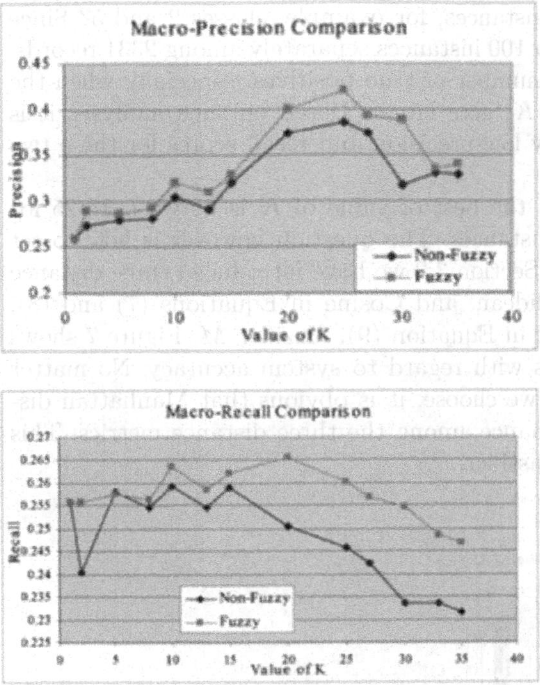
Macro-Precision Comparison
0.45
0.4
0.35
0.3
0.25
0.2
Precision
0 10 20 30 40
Value of K
Non-Fuzzy
Fuzzy
Macro-Recall Comparison
0.27
0.265
0.26
0.255
0.25
0.245
0.24
0.235
0.23
0.225
Recall
0 5 10 15 20 25 30 35 40
Value of K
Non-Fuzzy
Fuzzy

Fig. 6 Macro-precision and macro-recall comparison when Manhattan distance and Euclidean distance are applied respectively

Table 2 Class Distribution and Precision/Recall

Class ID	# of Records	Precision	Recall
0	253	0.40	0.38
1	162	0.36	0.25
2	98	0	0
3	158	0.33	0.28
4	230	0.33	0.17
5	92	0.25	0.01
6	414	0.66	0.54
7	632	0.37	0.90
8	292	0.58	0.49

about the classes with fewer instances, for example, classes 2 and 5? Since these two classes have less than 100 instances separately among 2331 records, this dramatically reduces the number of true positives, especially when the number of nearest neighbors, K, gets larger. Based on such analysis, it is not surprising to see extremely low precision and recall scores for these two classes.

Our experiments show that the best of value of K is between 15-25 for our yeast gene data of 2331 instances. The question however is how to set up other input parameters. In Section 3.2 we have introduced three distance metrics, i.e., Manhattan, Euclidean, and Cosine in Equations (7) and (8), and one more input parameter in Equation (9), fuzzifier M. Figure 7 shows the effect the distance metrics with regard to system accuracy. No matter which classification algorithm we choose, it is obvious that Manhattan distance renders the best performance among the three distance metrics. This is especially true with fuzzy classifier.

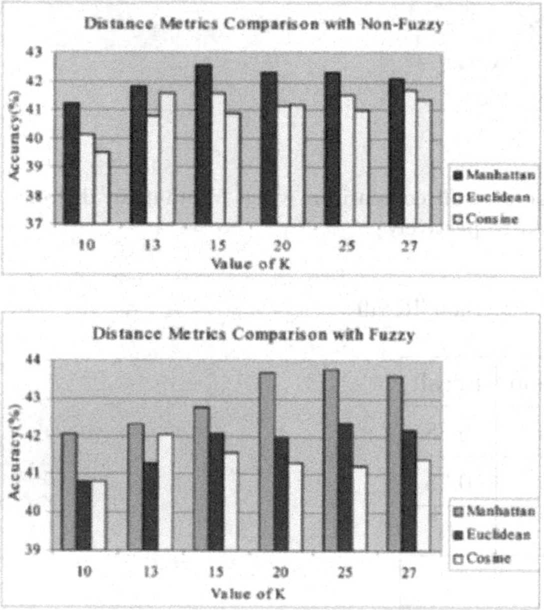
Distance Metrics Comparison with Non-Fuzzy
Accuracy(%)
Value of K
Manhattan
Euclidean
Cosine
Distance Metrics Comparison with Fuzzy
Accuracy(%)
Value of K
Manhattan
Euclidean
Cosine

Fig. 7 Distance Metrics Comparison for Crisp and Fuzzy Classifiers

Figure 8 shows the variation of system accuracy as the value of M changes in fuzzy classification. We can see that the best performance can be reached when M has a value between 1.3 and 1.5. This is consistent with the recommendation in [37], which suggested a value between 1 and 1.5 for fuzzifier.

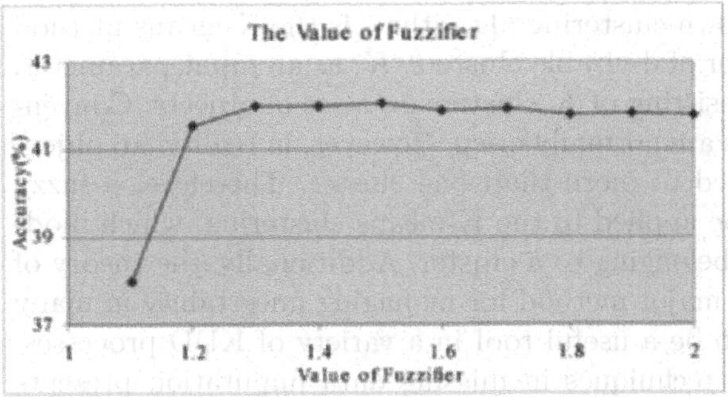
The Value of Fuzzifier
Accuracy(%)
43
41
39
37
1
1.2
1.4
1.6
1.8
2
Value of Fuzzifier

Fig. 8 The Effect of Fuzzifier

3.3 APPLICATION 3: Rough-Fuzzy K-Means Clustering for Missing Data Imputation

3.3.1 Missing Data Imputation — Problem Description

The problem of missing (or incomplete) data is relatively common in many fields, and it may have different causes such as equipment malfunction, unavailability of equipment, refusal of respondents to answer certain questions, etc. The overall result is that the observed data cannot be analyzed because of the incompleteness of the data. A number of researchers over the last several decades have investigated techniques for dealing with missing data [16, 25, 26, 27, 47, 51, 55, 56, 62]. Methods for handling missing data can be divided into three categories. The first is *ignoring and discarding data. Listwise deletion* and *pairwise deletion* are two widely used methods in this category [25]. The second group includes the methods based on *parameter estimation*, which uses variants of the *Expectation-Maximization* algorithm to estimate parameters in the presence of missing data [16]. The third category is *imputation*, which denotes the process of filling in the missing values in a data set by some plausible values based on information available in the data set [51].

Among imputation methods, there are many approaches varying from simple methods such as mean imputation, to some more robust and intricate methods based on the analysis of the relationships among attributes. For example, in hot deck imputation, the missing data are replaced by other cases with the same (or similar) characteristics. These common characteristics are derived from auxiliary variables, e.g., age, gender, race, or education degree, whose values are available from the cases to be imputed. Generally, there are two steps in hot deck imputation [24]. First, data are partitioned into several clusters based on certain similarity metric, and each instance with missing data is associated with one of the clusters. Second, by calculating the mean of the attribute within a cluster, the complete cases in the cluster are used to fill in the missing values.

One of the most well known clustering algorithms is the K-means method [30], which takes the number of desirable clusters, K, as an input parameter, and outputs a partition consisting of K clusters on a set of objects. Conventional clustering algorithms are normally crisp. However, in reality, an object sometimes could be assigned to more than one cluster. Therefore, a fuzzy membership function can be applied to the K-means clustering, which models the degree of an object belonging to a cluster. Additionally, the theory of rough set has emerged as a major method for managing uncertainty in many domains, and has proved to be a useful tool in a variety of KDD processes. The use of soft computing techniques in missing data imputation presents the major difference of our approach from that presented in [24].

3.3.2 Missing Data Imputation with K-Means Clustering

A fundamental problem in missing data imputation is to fill in missing information about an object based on the knowledge of other information about the object [63]. As one of the most popular techniques in data mining, the clustering method facilitates the process of solving this problem. Given a set of objects, the overall objective of clustering is to divide the data set into groups based on similarity of objects and to minimize the intra-cluster dissimilarity. In K-means clustering, the intra-cluster dissimilarity is measured by the summation of distances between the objects and the centroid of the cluster they are assigned to. A cluster centroid represents the mean value of the objects in a cluster. A number of different distance functions, e.g., Euclidean distance, Cosine-based distance, can be used.

Given a set of N objects $X = \{x_1, x_2, ..., x_N\}$ where each object has S attributes, we use x_{ij} ($1 \leq i \leq N$ and $1 \leq j \leq S$) to denote the value of attribute j in object x_i. Object x_i is called a *complete* object, if $\{x_{ij} \neq \phi \mid \forall\ 1 \leq j \leq S\}$, and an *incomplete* object, if $\{x_{ij} = \phi \mid \exists\ 1 \leq j \leq S\}$, and we say object x_i has a missing value on attribute j. For any incomplete object x_i, we use $R = \{j \mid x_{ij} \neq \phi, 1 \leq j \leq S\}$ to denote the set of attributes whose values are available, and these attributes are called *reference* attributes. Our objective is to obtain the values of non-reference attributes for the incomplete objects. By K-means clustering method, we divide dataset X into K clusters, and each cluster is represented by the centroid of the set of objects in the cluster. Let $V = \{v_1, v_2, ..., v_K\}$ be the set of K clusters, where v_k ($1 \leq k \leq K$) represents the centroid of cluster k. Note that v_k is also a vector in an S-dimensional space. We use $d(v_k, x_i)$ to denote the distance between centroid v_k and object x_i.

Figure 9 shows the algorithm for missing data imputation with K-means clustering method. The algorithm can be divided into three processes. First (Step 1), randomly select K complete data objects as K centroids . Rather than random selection, an alternative is to choose the first centroid as the object that is most central to the data set, and then pick other $(k-1)$ centroids one by one in such a way that each one is most dissimilar to all the objects

```
Algorithm: K-means-imputation(X, K, ε)
1) Initialization -- randomly select K complete objects from
X as centroids;
2) Assign each object (complete or incomplete) in X to the
closest cluster centroid;
3) Recompute the centroid of each cluster;
4) Repeat steps 2 & 3, until ∑_{k=1}^{K} ∑_{i=1}^{N} d(v_k, x_i) < ε;
5) For each incomplete object, apply nearest neighbor
algorithm to fill in all the non-reference attributes.
```

Fig. 9 K-means Clustering for Missing Data Imputation

that have already been selected. This makes the initial K centroids evenly distributed. Second (Steps 2 to 4), iteratively modify the partition to reduce the sum of the distances for each object from the centroid of the cluster to which the object belongs. The process terminates once the summation of distances is less than a user-specified threshold ε. The last process (Step 5) is to fill in all the non-reference attributes for each incomplete object based on the cluster information. Data objects that belong to the same cluster are taken as nearest neighbors of each other. The missing data are replace by Inverse Distance Weighted (IDW) approach based on the available data values from nearest neighbors.

Generalized L_P norm distance and Cosine-based distance (as described in Section 3.2) are used as distance metrics. The distance functions are normalized for two reasons. First, the distances can be calculated only from the values of reference attributes, but for incomplete objects, the number of reference attributes is different. Second, each attribute (either numerical or categorical) has a different domain and the distance functions do not make sense without normalization. Because the domain of each attribute is already known in our application domains, we employ the min-max method to normalize the input data sets.

3.3.3 Missing Data Imputation with Fuzzy K-Means Clustering

Now, the original K-means clustering method is extended to a fuzzy version to impute missing data. The reason for applying the fuzzy approach is that fuzzy clustering provides a better tool when the clusters are not well-separated, as is sometimes the case in missing data imputation. Moreover, the original K-means clustering may be trapped in local minimum if the initial points are not selected properly. However, continuous membership values in fuzzy clustering make the resulting algorithms less susceptible to get stuck in local minimum [35].

In fuzzy clustering, each data object x_i has a membership function which describes the degree that this data object belongs to certain cluster v_k. The membership function is:

$$U(v_k, x_i) = \frac{d(v_k, x_i)^{-2/(m-1)}}{\sum_{j=1}^{K} d(v_j, x_i)^{-2/(m-1)}}, \tag{10}$$

where $m > 1$ is the fuzzifier and $\sum_{j=1}^{K} U(v_j, x_i) = 1$ for any data object x_i $(1 \leq i \leq N)$ [38]. Now, the cluster centroids cannot be calculated simply by the mean values. Instead, the calculation a each cluster centroid needs to consider the membership degree of each data object. The formula for cluster centroid computation is:

$$v_k = \frac{\sum_{i=1}^{N} U(v_k, x_i) * x_i}{\sum_{i=1}^{N} U(v_k, x_i)}. \tag{11}$$

Because there are unavailable data in incomplete objects, the fuzzy K-means approach uses only reference attributes to compute the cluster centroids.

Figure 10 shows the algorithm for missing data imputation with fuzzy K-means clustering method. This algorithm has three processes which are the same as *K-means-imputation*. In the initialization process (Steps 1 & 2), the algorithm picks K centroids which are evenly distributed to avoid local minimum situation. The second process (Steps 3 to 5), iteratively updated membership functions and centroids until the overall distance meets the user-specified distance threshold ε. In this process, a data object cannot be assigned to a concrete cluster represented by a cluster centroid (as did in the basic K-mean clustering algorithm), because each data object belongs to all K clusters with different membership degrees. Finally (Step 6), the algorithm replaces non-reference attributes for each incomplete object.

Algorithm *K-means-imputation* fills in missing data by a nearest neighbor algorithm which takes the data points belonging to the same cluster as nearest neighbors. However, in *Fuzzy-K-means-imputation*, the nearest neighbors are not available, because clusters are not well-separated with regard to the fuzzy concept. *Fuzzy-K-means-imputation* replaces non-reference attributes for each incomplete data object x_i based on the information about membership degrees and the values of cluster centroids,

$$x_{i,j} = \sum_{k=1}^{K} U(x_i, v_k) * v_{k,j}, \text{ for any non-refence attribute } j \notin R. \tag{12}$$

Algorithm: `Fuzzy-K-means-imputation(X, K, ε)`
1) Compute the most centered complete object and select it as the first centroid, i.e.,

$$v_1 = \min_{1 \le i \le N} \sum_{j=1}^{N} d(x_i, x_j);$$

2) Select other $(K - 1)$ complete objects as centroids such that each one is most dissimilar to all the centroids that have already been selected, i.e.,

$$\text{for } (2 \le i \le K) \ \{v_i = \max_{1 \le j \le N, x_j \notin V} (\min_{1 \le k \le K, v_k \in V} d(x_j, v_k))\};$$

3) Compute the membership function $U(v_k, x_i)$ using Equation (10) for each $1 \le k \le K$, and $1 \le i \le N$;
4) Recompute centroid v_k using Equation (11);
5) Repeat steps 3 & 4, until $\sum_{k=1}^{K} \sum_{i=1}^{N} U(v_k, x_i) d(v_k, x_i) < \varepsilon$;
6) Fill in all the non-reference attributes for each incomplete data object.

Fig. 10 Fuzzy K-means Clustering for Missing Data Imputation

3.3.4 Missing Data Imputation with Rough K-Means Clustering

This section presents a missing data imputation algorithm based on rough set theory. Theories of rough set and fuzzy set are distinct generalizations of set theory [52, 64]. A fuzzy set allows a membership value between 0 and 1 which describes the degree that an object belongs to a set. Based on rough set theory, a pair of upper and lower bound approximations are used to describe a reference set. Given an arbitrary set X, the lower bound $\underline{A}(X)$ is the union of all elementary sets, which are subsets of X. The upper bound $\overline{A}(X)$ is the union of all elementary sets that have a non-empty intersection with X [52]. In other words, elements in the lower bound of X definitely belong to X, while elements in the upper bound of X may or may not belong to X.

In the original crisp K-means clustering algorithm, data objects are grouped into the same cluster if they are close to each other and each data object belongs to only one cluster. In the rough K-means algorithm, each cluster is represented by two sets which include all the data objects that approximate its lower bound and upper bound, respectively. Different from crisp K-means method, in rough K-means, a data object may exist in the upper bound of one or more clusters. One of the most important issues in the rough K-means clustering is how to assign each data object into the lower or upper bound of one or more clusters. In the crisp K-means algorithm, data objects are assigned to different clusters simply based on the distances between data objects and cluster centroids. The rough K-means clustering still uses distance metrics defined earlier to determine cluster membership, but the process is more complicated because each cluster is represented by both the lower and upper bound approximations. This process is shown in Figure 11. A new parameter, θ, is introduced which is used to control the similarity among the data objects belonging to a common upper bound of a cluster. Algorithm *Rough-Assignment* shows that in rough K-means clustering, each data object can only belong to the lower bound of one cluster, but it may exist in the upper bound of one or more clusters.

Algorithm: `Rough-Assignment` (x_i, K, θ)
1) Find the cluster centroid v_k to which the data object x_i has the minimum distance, i.e. $v_k = \min\ d(v_{k'}, x_i)$ for all $1 \leq k' \leq K$;
2) x_i is assigned to the lower and upper bounds of cluster v_k, i.e. $x_i \in \underline{A}(v_k)$ and $x_i \in \overline{A}(v_k)$, if $d(v_{k'}, x_i) - d(v_k, x_i) > \theta$ for all $1 \leq k' \leq K$, and $k' \neq k$; otherwise
3) x_i is assigned to the upper bounds of clusters v_k and $v_{k'}$, i.e. $x_i \in \overline{A}(v_k)$ and $x_i \in \overline{A}(v_{k'})$, if $d(v_{k'}, x_i) - d(v_k, x_i) \leq \theta$ for any $1 \leq k' \leq K$, and $k' \neq k$.

Fig. 11 Data Object Assignment in Rough K-means algorithm

Another important modification in rough K-means clustering is the computation of cluster centroids. Each cluster is represented by two sets, the lower bound approximation and the upper bound approximation. Both sets are used to re-compute the value of a cluster centroid [45]:

$$v_k = \begin{cases} \dfrac{\sum\limits_{x_i \in \underline{A}(v_k)} x_i}{|\underline{A}(v_k)|} \times W_{lower} + \dfrac{\sum\limits_{x_i \in (\overline{A}(v_k) - \underline{A}(v_k))} x_i}{|\overline{A}(v_k) - \underline{A}(v_k)|} \times W_{upper}, & \text{if } |\overline{A}(v_k)| \neq |\underline{A}(v_k)|, \\[2ex] \dfrac{\sum\limits_{x_i \in \underline{A}(v_k)} x_i}{|\underline{A}(v_k)|}, & \text{otherwise.} \end{cases} \qquad (13)$$

In this equation, there are two more parameters, W_{lower} and W_{upper}, which are used to control the relative importance of lower and upper bound approximations. For the purpose of normalization, the equation does not use the weight function in the second case. This is different from the equation given in [45]. Generally, $W_{lower} + W_{upper} = 1$ and $W_{lower} \geq W_{upper}$, based on the definitions of lower and upper bounds in rough set theory. If a cluster includes an incomplete data object, only the reference attributes of the data object are used for centroid computation.

Overall, the major difference between rough K-means and crisp K-means imputation methods lies in the second process. For rough imputation algorithm, each data object is assigned to the lower or upper bound of one or more clusters based on *Rough-Assignment* process and re-computed the centroid for each cluster based on Equation (13).

The imputation methods applied to crisp K-means and fuzzy K-means clusterings cannot be applied to rough K-means clustering. In crisp K-means clustering, a data object only belongs to one cluster, and in fuzzy K-means clustering, a data object belongs to all K clusters with different membership degrees. However, in rough K-means clustering, an incomplete data object either exists in the lower bound of one cluster (also in the upper bound of this cluster) or exists in the upper bounds of two or more clusters. Equation (14) shows how we deal with these two different situations:

$$x_i = \begin{cases} \dfrac{\sum\limits_{x_j \in \underline{A}(v_k)} x_j}{|x_j|} \times W_{lower} + \dfrac{\sum\limits_{x_j \in (\overline{A}(v_k) - \underline{A}(v_k))} x_j}{|x_j|} \times W_{upper}, \\ \text{if } x_i \in \underline{A}(v_k) \text{ for any } 1 \le k \le K, \text{ and } x_j \text{ is a complete object,} \\ \dfrac{\sum\limits_{x_j \in \overline{A}(v_k)} x_j}{|x_j|}, \text{ if } x_i \notin \underline{A}(v_{k'}) \text{ for all } 1 \le k' \le K. \end{cases} \tag{14}$$

For fuzzy K-means imputation, the computation of a non-reference attribute is based on the values of cluster centroids and the information about membership degrees. This is feasible because each cluster includes all data objects, and the cluster centroids, in turn, are calculated based on all data points. For rough K-means imputation, to make the algorithm more accurate, the value of an incomplete data object is computed based on the values of data objects (rather than cluster centroids) that are in the same cluster as the imputed data object. Moreover, two weight parameters, W_{lower} and W_{upper}, are used if the imputed data object exists in both the lower and upper bounds of a cluster.

3.3.5 Missing Data Imputation with Rough-Fuzzy K-Means Clustering

There are ongoing efforts to integrate fuzzy logic with rough set theory for dealing with uncertainty arising from inexact or incomplete information [6, 8, 35]. In this section, we present a rough-fuzzy hybridization method to capture the intrinsic uncertainty involved in cluster analysis. In this hybridization, fuzzy sets help handle ambiguity in input data, while rough sets

Algorithm: *Rough-Fuzzy-Assignment* (x_i, K, θ)
1) Find the cluster centroid v_k to which the data object x_i has the minimum distance, i.e. $v_k = \min d(v_{k'}, x_i)$ for all $1 \leq k' \leq K$;
2) Assign x_i to the lower bound of cluster v_k, i.e. $x_i \in \underline{A}(v_k)$;
3) Assign x_i to the lower bounds of clusters $v_{k'}$, i.e. $x_i \in \underline{A}(v_{k'})$, if there exists $1 \leq k' \leq K$, and $k' \neq k$ such that $d(v_{k'}, x_i) - d(v_k, x_i) \leq \theta$;
4) Assign x_i to the upper bound of each cluster, i.e. $x_i \in \overline{A}(v_k)$, for all $1 \leq k \leq K$.

Fig. 12 Data Object Assignment in Rough-Fuzzy Algorithm

represent each cluster with lower and upper approximations. In rough K-means clustering, a data object either exists in the lower bound of one cluster or exists in the upper bounds of two or more clusters. To deal with the uncertainty involved in lower and upper bound approximations,the rough K-means clustering assigns a data object to the lower bounds of two or more clusters. At the same time, each data object belongs to the upper bounds of all clusters with different membership degrees. This drives the main idea of the rough-fuzzy K-means clustering algorithm.

Figure 12 shows the algorithm for data object assignment in rough-fuzzy clustering . From the description of the algorithm, each data object may be assigned to the lower bound of one or more clusters depending on the value of distance, and each object is assigned to the upper bound of every cluster. Therefore, each data object x_i has two membership functions which describe the degrees that this data object belongs to the lower and upper bounds of certain cluster v_k. The membership functions are defined in Equations (15) and (16). Here, $\underline{U}(v_k, x_i)$ denotes the membership degree that data object x_i belongs to the lower bound of cluster v_k ($\underline{U}(v_k, x_i) = 0$, if $x_i \notin \underline{A}(v_k)$), and $\overline{U}(v_k, x_i)$ denotes the membership degree that data object x_i belongs to the upper bound of cluster v_k. $\sum_{k=1}^{K} \underline{U}(v_k, x_i) = \sum_{k=1}^{K} \overline{U}(v_k, x_i) = 1$ for any data object x_i ($1 \leq i \leq N$).

$$\underline{U}(v_k, x_i) = \frac{d(v_k, x_i)^{-2/(m-1)}}{\sum\limits_{x_i \in \underline{A}(v_j)} d(v_j, x_i)^{-2/(m-1)}}. \tag{15}$$

$$\overline{U}(v_k, x_i) = \frac{d(v_k, x_i)^{-2/(m-1)}}{\sum\limits_{j=1}^{K} d(v_j, x_i)^{-2/(m-1)}}. \tag{16}$$

To accommodate the properties of fuzzy and rough sets, we combine Equations (11) and (13) into a new formula to calculate cluster centroids in rough-fuzzy clustering algorithm, as shown in Equations (17).

$$v_k = \begin{cases} \dfrac{\sum\limits_{x_i \in \underline{A}(v_k)} \underline{U}(v_k, x_i) * x_i}{\sum\limits_{x_i \in \underline{A}(v_k)} \underline{U}(v_k, x_i)} \times W_{lower} + \dfrac{\sum\limits_{x_i \notin \underline{A}(v_k)} \overline{U}(v_k, x_i) * x_i}{\sum\limits_{x_i \notin \underline{A}(v_k)} \overline{U}(v_k, x_i)} \times W_{upper}, \\ \qquad\qquad\qquad\qquad\qquad\qquad \text{if } |\overline{A}(v_k)| \neq |\underline{A}(v_k)|; \\ \dfrac{\sum\limits_{i=1}^{N} \overline{U}(v_k, x_i) * x_i}{\sum\limits_{i=1}^{N} \overline{U}(v_k, x_i)}, \qquad\qquad \text{otherwise.} \end{cases} \tag{17}$$

The computation of a non-reference attribute for an incomplete data object is based on two parts considering both lower and upper approximations of a cluster, as shown in Equation (18). Because $\sum_{k=1}^{K} \underline{U}(v_k, x_i) = \sum_{k=1}^{K} \overline{U}(v_k, x_i) = 1$ and $W_{lower} + W_{upper} = 1$, this computation formula is well normalized.

$$x_i = \sum_{k=1,\ x_i \in \underline{A}(v_k)}^{K} \frac{\sum_{j=1,\ x_j \in \underline{A}(v_k)}^{N} \underline{U}(v_k, x_i) * x_j}{|x_j|} \times W_{lower} + \sum_{k=1}^{K} \frac{\sum_{j=1,\ x_j \notin \underline{A}(v_k)}^{N} \overline{U}(v_k, x_i) * x_j}{|x_j|} \times W_{upper}. \tag{18}$$

The rough-fuzzy K-means imputation algorithm is shown in Figure 13.

Algorithm: ***Rough-Fuzzy-K-means-imputation(X, K, ε)***

1) Select K initial data objects as cluster centroids;
2) Assign each data object x_i in X to the appropriate lower and upper bounds with Algorithm *Rough-Fuzzy-Assignment*;
3) Compute the membership functions $\underline{U}(v_k, x_i)$ and $\overline{U}(v_k, x_i)$ using Equation (8) and (9) for each $1 \leq k \leq K$;
4) Recompute cluster centroid v_k using Equation (10);
5) Repeat steps 2, 3 & 4, until distance threshold ε is satisfied;
6) Fill in all the non-reference attributes using Equation (11) for each incomplete data object.

Fig. 13 Rough-Fuzzy Clustering for Missing Data Imputation

3.3.6 Experiments and Analysis

Two types of experiments are designed. First, the algorithms are evaluated based on complete datasets which are subsets of real-life databases without incomplete data objects. The overall objective of the experiments is to find the best value for each of the parameters (e.g,. missing percentage, the fuzzifier value, and the number of clusters, etc.). Second, the algorithms are evaluated based on real-life datasets with missing values. The best parameter values discovered earlier are used in this process. There are two types of real-life datasets. One is weather databases for drought risk management. Weather data are collected at automated weather stations in Nebraska. These weather stations serve as long-term reference sites to search for key patterns among climatic events. The other type of data is the Integrated Psychological Therapy (IPT) outcome databases for psychotherapy study. A common property in these two types of datasets is that missing data are present either due to the malfunction (or unavailability) of equipment or caused by the refusal of respondents. The experimental results shown in this section are based on the monthly weather data in Clay Center, NE, from 1950-1999. The dataset includes ten fields. Because each data attribute has different domain, to test the algorithms meaningfully, the dataset is first normalized so that all the data values are between 0 and 100. The experiments are based on the following input parameters: distance metric = Manhattan distance; the number of cluster $K = 7$; the fuzzifier $m = 1.2$; the percentage of missing data = 5%; the distance threshold $\theta = 1$; $W_{lower} = 0.9$; and $W_{upper} = 0.1$.

We evaluate the quality of algorithms based on cross validation resampling method. Each algorithm is tested ten times and each time a sample is randomly divided into two subsets, test set and training set. The test results are validated by comparing across sub-samples. The Root Mean Squared Error (RMSE) is selected to compare the prediction value with the actual value of a test instance. RMSE error analysis metric is defined as follows:

$$RMSE = \sqrt{\frac{\sum_{i=1}^{n} |F_i - f_i|^2}{n}},$$

where n is the total number of test points, F_i are the estimated data values, and f_i are the actual data values. Note that the RMSE is much biased because it exaggerates the prediction error of test cases in which the prediction error is larger than others. However, from another point of view, if the RMSE number is significantly greater than zero, it means that there are test cases in which the prediction value is significantly greater or less than the actual value. Therefore, sensitivity of RMSE number is useful in highlighting test cases in which prediction value is significantly lower or higher.

1. Experiments on Complete Datasets
The performance of the four K-means imputation algorithms is evaluated and analyzed from two aspects. First, the experiments show the influence of the missing percentage. Second, the experiments test various input parameters (i.e., distance metrics, the value of fuzzifier m, and cluster number K, etc.), and conclude with the best values. The evaluation of these two aspects is based on complete datasets, which are subsets of real-life datasets without incomplete data objects.

Percentage of Missing Data. Figure 14 summarizes the results for varying percentages of missing values in the test cases. Besides the four K-means imputation algorithms, the experiments also test a widely used missing data imputation algorithm, mean substitution. There are four observations from Figure 14:

1. As the percentage of missing values increases, the overall error also increases considering all of these five algorithms. This is reasonable because we lose more useful information when the amount of missing data increases.
2. When the missing percentage is less than or equal to 40%, rough-fuzzy K-means algorithm provides the best results, while the performance of mean substitution imputation algorithm is the worst.
3. When the missing percentage is greater than 15%, the curve for the crisp K-means algorithm terminates, as shown in Figure 14. This occurs because for any incomplete data object, when filling in the values for its non-reference attributes, the algorithm needs to have the values on these attributes from other data objects which are within the same cluster as this incomplete object. However, it is possible that all the data objects within the same cluster have a common non-reference attribute. In this case, the nearest neighbor algorithm used for K-means imputation will

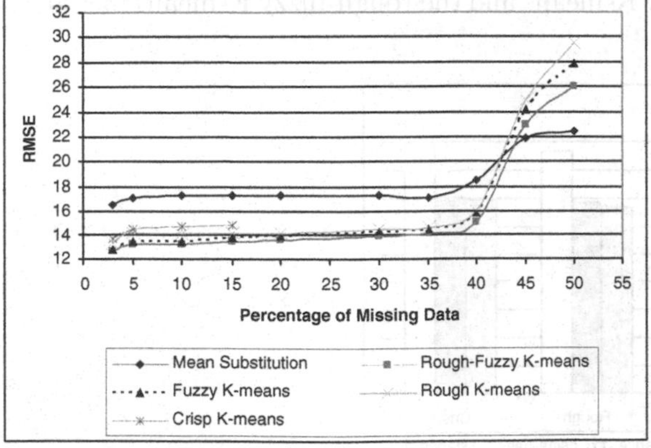
RMSE
32
30
28
26
24
22
20
18
16
14
12
0
5
10
15
20
25
30
35
40
45
50
55
Percentage of Missing Data
Mean Substitution
Fuzzy K-means
Crisp K-means
Rough-Fuzzy K-means
Rough K-means

Fig. 14 RMSE for Varying Percentages of Missing Values

not work. This will not happen in fuzzy or rough algorithms. In the fuzzy imputation algorithm, the final imputation process is based on the centroid information and the membership degrees. These two kinds of information are always available for computation. In the rough imputation algorithm, an incomplete data object may belong to two or more clusters, and the information on lower and upper bounds for a given cluster makes the computation flexible and feasible to deal with uncertainty.

4. There is a sharp increase in the value of RMSE when the missing percentage is greater than 40% considering the four K-means imputation methods. The mean substitution approach outperforms the four imputation algorithms when the missing percentage is greater than 45%. This indicates that the four K-means algorithms cannot properly discover the similarity among data objects when there are too many missing values.

Distance Metrics. The experiments are designed to evaluate the four missing data imputation algorithms by testing on different input parameters. First, the experiments test three distance metrics, Euclidean distance, Manhattan distance, and Cosine-based distance, as shown in Equations 7 and 8. Figure 15 presents the influence of these metrics. The performance of the four imputation algorithms is shown in four different groups. Considering all of these four algorithms, Manhattan distance provides the best performance while the Cosine-based distance metric is the worst.

Values of Fuzzifier in Fuzzy Algorithms. The experiments test the effect of the value of fuzzifier in the fuzzy and the rough-fuzzy K-means imputation methods. Because fuzzifier is a parameter only in the fuzzy imputation algorithms, as shown in Figure 16, the RMSE in the crisp K-means and the rough K-means clustering methods does not change much as the value of m changes. However, for the fuzzy algorithms, the change in performance is obvious, and the best value of m is 1.2 for both the fuzzy and the rough-fuzzy algorithms. When the value of fuzzifier goes to 1.5, the crisp K-means algorithm outperforms the fuzzy K-means and the rough-fuzzy K-means methods.

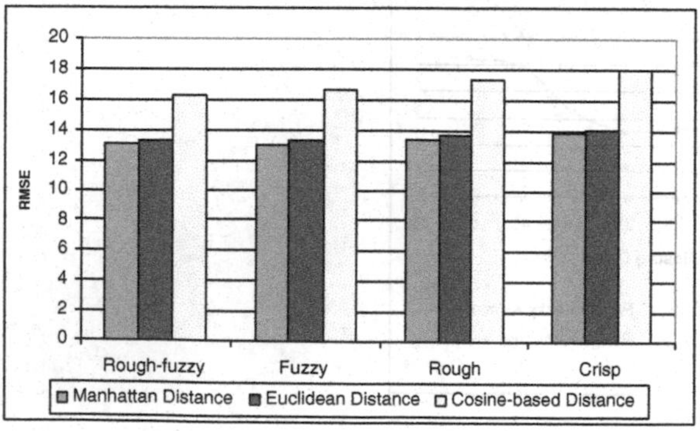
20
18
16
14
12
10
8
6
4
2
0
RMSE
Rough-fuzzy
Fuzzy
Rough
Crisp
Manhattan Distance
Euclidean Distance
Cosine-based Distance

Fig. 15 RMSE for Varying Distance Metrics

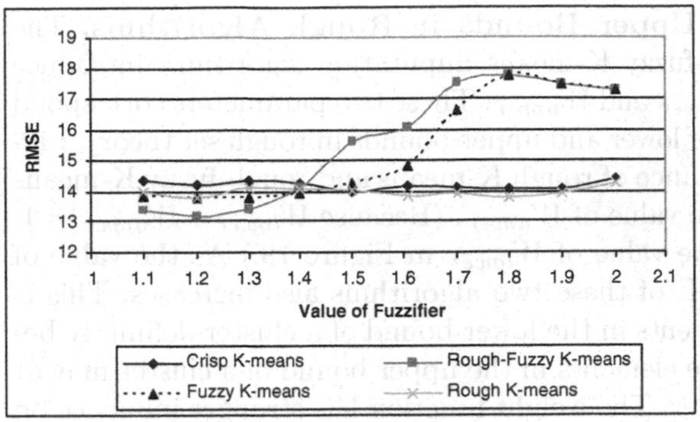
RMSE
19
18
17
16
15
14
13
12
1
1.1
1.2
1.3
1.4
1.5
1.6
1.7
1.8
1.9
2
2.1
Value of Fuzzifier
Crisp K-means
Rough-Fuzzy K-means
Fuzzy K-means
Rough K-means

Fig. 16 RMSE for Varying the Value of Fuzzifier

This indicates that selecting a proper parameter value is important for system performance. Moreover, the experimental results are consistent with the recommendation in [38], which suggested a value between 1 and 1.5 for m.

Number of Clusters. Now, the experiments test the influence of the number of clusters, K. The value of K is varied from 4 to 11. Figure 17 shows the performance of the algorithms when there are 6000 data items in the test dataset. From the figure, the best value of K is 7 for all four algorithms. It is worth mentioning that for $K = 4$, the crisp K-means algorithm is the best one among all four algorithms. This is because the smaller number of clusters have fewer centroids. This, in turn, limits the possible variance in the imputed data values for the other three K-means imputation algorithms. On the other hand, when the number of clusters is small, the number of data objects in each cluster increases. This provides more information for the basic K-means algorithm when nearest neighbor algorithm is applied to estimate missing values.

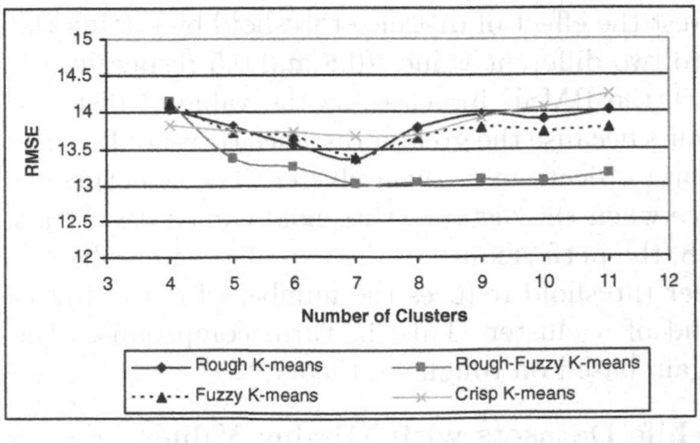
RMSE
15
14.5
14
13.5
13
12.5
12
3
4
5
6
7
8
9
10
11
12
Number of Clusters
Rough K-means
Rough-Fuzzy K-means
Fuzzy K-means
Crisp K-means

Fig. 17 RMSE for Varying the Number of Clusters

Weights of Lower and Upper Bounds in Rough Algorithms. The rough K-means and rough-fuzzy K-means imputation algorithms introduce two weight parameters, W_{lower} and W_{upper}. These two parameters correspond to the relative importance of lower and upper bounds in rough set theory. Figure 18 presents the performance of rough K-means and rough-fuzzy K-means algorithms as we change the value of W_{upper}. (Because $W_{lower} + W_{upper} = 1$, Figure 18 does not show the value of W_{lower} in Figure 18.) As the value of W_{upper} increases, the RMSE of these two algorithms also increases. This is reasonable because the elements in the lower bound of a cluster definitely belong to the cluster, while the elements in the upper bound of a cluster may or may not belong to the cluster. The weight function has stronger influence on the rough K-means imputation algorithm than on the rough-fuzzy K-means algorithm, because fuzzy sets help handle ambiguity in cluster information.

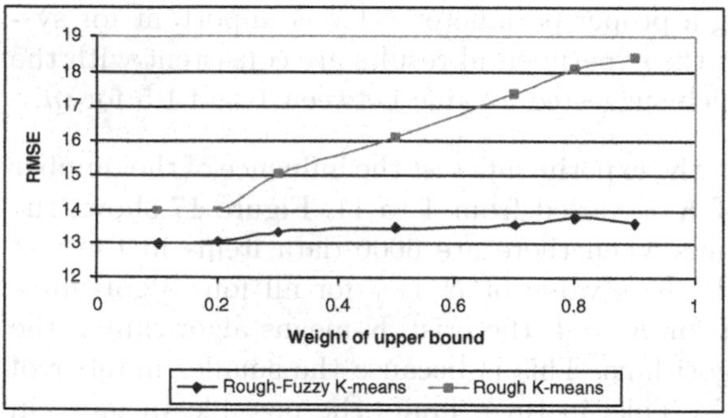
RMSE
19
18
17
16
15
14
13
12
0
0.2
0.4
0.6
0.8
1
Weight of upper bound
Rough-Fuzzy K-means
Rough K-means

Fig. 18 RMSE for Varying the Weight of Lower and Upper Bounds

Distance Threshold in Rough Algorithms. The rough K-means and rough-fuzzy K-means algorithms use distance threshold to control the similarity between data objects that belong to the same upper (for rough imputation algorithms) or lower (for rough-fuzzy imputation algorithm) bound of a cluster. The experiments test the effect of distance threshold by setting the weight of the lower bound to two different values (0.8 and 0.5 respectively). As can be seen in Figure 19, the RMSE increases as the value of distance threshold increases. This occurs because the greater distance threshold results in less similarity between data objects in a given cluster. The experiments present the best performance when the distance threshold equals 0.8. When the threshold is less than 0.8, the performance of the two algorithms slightly decreases because the smaller threshold reduces the number of data objects in the upper or lower bound of a cluster. This, in turn, compromises the possible benefit we should gain based on rough set theory.

2. Experiments on Real-Life Datasets with Missing Values

The previous experiments are based on test datasets, which are subsets of real-life datasets without incomplete data objects. However, in reality, the

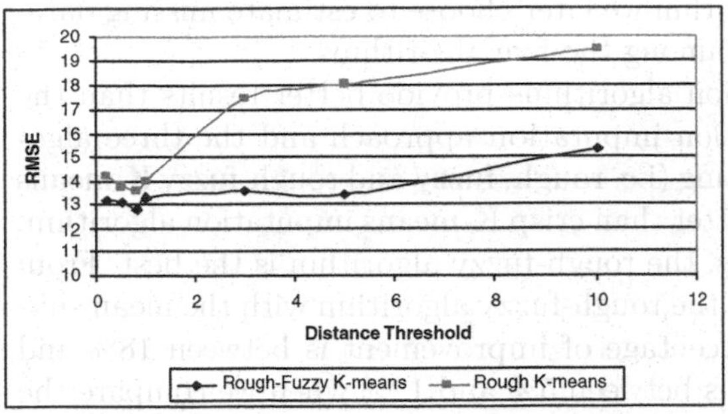
RMSE
Distance Threshold
Rough-Fuzzy K-means
Rough K-means

(a) When the weight of lower bound equals 0.8.

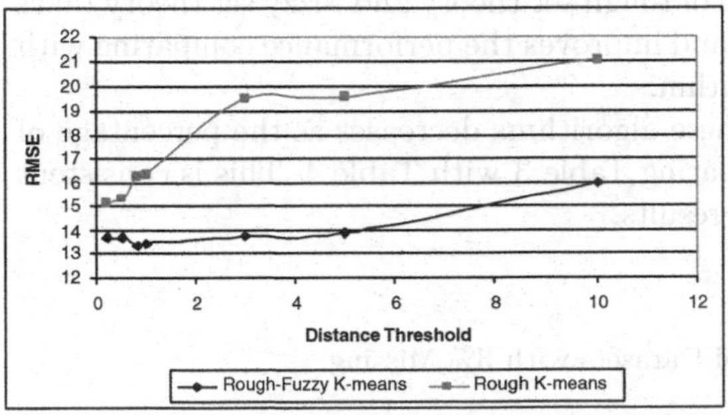
RMSE
Distance Threshold
Rough-Fuzzy K-means
Rough K-means

(b) When the weight of lower bound equals 0.5.

Fig. 19 RMSE for Varying the Value of Distance Threshold

datasets include incomplete data objects, and we do not have actual data values on the non-reference attributes for an incomplete data object. Therefore, the algorithms cannot be evaluated simply based on the root mean square error. To solve this problem, the experiments are designed in this way: 1) Initially fill the missing data with one of the algorithms and get a complete dataset. 2) From this new dataset, randomly remove a certain percentage data which have actual data values in original dataset. 3) Different imputation algorithms are applied to estimate these missing data values. 4) The RMSE is computed for each algorithm based on estimated data values and actual data values.

In addition to the four K-means imputation algorithms, the mean substitution method is also implemented. Tables 3 and 4 compare the five algorithms when the percentage of missing data is 3% and 10% respectively. We make three observations from these two tables:

1. The experimental results are mainly determined by the algorithm which we initially select to estimate the actual missing data. This explains why the root mean square errors in each column in Table 3 and 4 have similar values. Based on this observation, once we fix the algorithm which is initially

used, no matter which algorithm we later choose to estimate missing data, there is no much difference among the five algorithms.
2. The four K-means imputation algorithms provide better results than the widely used mean substitution imputation approach and the three algorithms based on soft computing (i.e. rough, fuzzy and rough-fuzzy K-means imputation methods) are better than crisp K-means imputation algorithm. Among the three algorithms, the rough-fuzzy algorithm is the best. From the experiments, comparing the rough-fuzzy algorithm with the mean substitution algorithm, the percentage of improvement is between 18% and 27%, and the improvement is between 5% and 12% when we compare the rough-fuzzy imputation algorithm with the crisp K-means algorithm. This shows that the hybridization of rough set theory and fuzzy set theory takes advantages of both theories and improves the performance comparing with simple fuzzy or rough algorithm.
3. The performance of all of these algorithms decreases as the percentage of missing data increases comparing Table 3 with Table 4. This is consistent with previous experimental results.

Table 3 Experiments on Actual Datasets with 3% Missing

Fill actual data with	Initially fill missing data with				
	Mean Sub	Crisp K-m	Rough K-m	Fuzzy K-m	R-F K-m
Mean Substitution	16.22	13.72	13.48	13.45	13.14
Crisp K-means	16.43	13.35	12.92	12.81	12.57
Rough K-means	16.61	13.27	12.51	12.26	12.13
Fuzzy K-means	16.70	13.39	12.40	12.16	12.07
R-F K-means	16.65	13.35	12.29	12.14	12.04

Table 4 Experiments on Actual Datasets with 10% Missing

Fill actual data with	Initially fill missing data with				
	Mean Sub	Crisp K-m	Rough K-m	Fuzzy K-m	R-F K-m
Mean Substitution	17.19	14.43	13.57	13.53	13.31
Crisp K-means	17.24	14.15	13.42	13.21	13.03
Rough K-means	17.34	14.11	13.23	13.03	12.87
Fuzzy K-means	17.29	14.21	13.24	12.95	12.86
R-F K-means	17.30	14.32	13.27	12.94	12.85

4 Summary

Current research in data mining mainly focuses on the discovery of algorithms and visualization techniques. There is a growing awareness that, in practice, it is easy to discover a huge number of pattern in a database where some of these patterns are actually redundant, useless, or uninteresting to the user due to the uncertain nature of existing data. To prevent the user from being overwhelmed by a large number of uninteresting patterns, techniques are needed to identify only the useful/interesting patterns.

Soft computing differs from conventional (hard) computing in that, unlike hard computing, it is more tolerant of imprecision, uncertainty, partial truth, and approximation. Soft computing methodologies, involving fuzzy sets, neural networks, rough sets, and their hybridizations, have recently been used to solve data mining problems. They strive to provide approximate solutions at low cost, thereby speeding up the process [50]. This chapter presents three real-world projects which apply soft computing methodologies to existing data mining solutions and successfully handle uncertainty, imprecision, and uncertainty in data.

In summary, data mining is a practical problem that drives theoretical studies toward understanding and reasoning about large and existing data. While progress has been made in the direction of automatically acquiring knowledge needed for guiding and controlling the knowledge discovery process, the ideal system remains far from reach. At the system level, more research is needed in how to derive domain knowledge from databases and how to represent domain knowledge and derived knowledge in a uniform manner. At the level of methods for extracting patterns, we believe that data mining is an important application area where the theoretical results of fuzzy set theory and rough set theory can be tested, in order to help us understand its strengths and weaknesses.

References

[1] The Brown Lab, `http://brownlab.stanford.edu/`

[2] Munich information centre for protein sequence, `http://mips.gsf.de/proj/yeast/catalogues/funcat/`

[3] Agrawal, R., Imielinski, T., Swami, A.: Mining association rules between sets of items in large databases. In: Proceedings of the ACM SIGMOD 1993 International Conference on Management of Data [SIGMOD 1993], Washington D.C., pp. 207–216 (1993)

[4] Akleman, E., Chen, J.: Generalized distance functions. In: Proceedings of the 1999 International Conference on Shape Modeling, pp. 72–79 (March 1999)

[5] Altschul, S.F., Madden, T.L., Schaffer, A.A., Zhang, J., Zhang, Z., Miller, W., Lipman, D.J.: Gapped blast and psi-blast: a new generation of protein database search programs. Nucleic Acids Research (25), 3389–3402 (1997)

[6] Asharaf, S., Narasimha Murty, M.: An adaptive rough fuzzy single pass algorithm for clustering large data sets. Pattern Recognition 36, 3015–3018 (2003)
[7] Bace, R.: Intrusion Detection. Macmillan Technical Publishing, Basingstoke (2000)
[8] Banerjee, M., Mitra, S., Pal, S.K.: Rough fuzzy mlp: Knowledge encoding and classification. IEEE Trans. Neural Networks 9, 1203–1216 (1998)
[9] Barbara, D., Couto, J., Jajodia, S., Popyack, L., Wu, N.: ADAM: Detecting intrusions by data mining. In: Proc. of the 2001 IEEE Workshop on Information Assurance and Security, West Point, NY, pp. 11–16 (June 2001)
[10] Barbara, D., Couto, J., Jajodia, S., Wu, N.: ADAM: a testbed for exploring the use of data mining in intrusion detection. ACM SIGMOD Special Issue: Special section on data mining for intrusion detection and threat analysis 30(4), 15–24 (2001)
[11] Bondugula, R., Duzlevski, O., Xu, D.: Profiles and fuzzy k-nearest neighbor algorithm for protein secondary structure prediction. In: Proc. of the 3rd Asia-Pacific Bioinformatics Conference, Singapore, pp. 85–94 (January 2005)
[12] Cai, Y., Bork, P.: Homology-based gene prediction using neural nets. Anal. Biochem. (265), 269–274 (1998)
[13] Chan, K.C.C., Wong, A.K.C.: A statistical technique for extracting classificatory knowledge from databases. Knowledge Discovery in Databases, 107–124 (1991)
[14] Cho, H., Dhillon, I.S., Guan, Y., Sra, S.: Minimum sum-squared residue co-clustering of gene expression data. In: Proc. of the Fourth SIAM International Conference on Data Mining, Florida (2004)
[15] Corinna, C., Drucker, H., Hoover, D., Vapnik, V.: Capacity and complexity control in predicting the spread between barrowing and lending interest rates. In: Proceedings of the 1st International Conference on Knowledge Discovery and Data Mining, Montreal, Quebec, Canada, pp. 51–76 (1995)
[16] Dempster, A.P., Laird, N.M., Rubin, D.B.: Maximum likelihood from incomplete data via the EM algorithm. Journal of the Royal Statistical Society Series B 39(1), 1–38 (1977)
[17] Deogun, J., Raghavan, V., Sarkar, A., Sever, H.: Data mining: Trends in research and development. Rough Sets and Data Mining: Analysis for Imprecise Data, 9–45 (1996)
[18] Dokas, P., Ertoz, L., Kumar, V., Lazarevic, A., Srivastava, J., Tan, P.: Data mining for network intrusion detection. In: Proceedings of NSF Workshop on Next Generation Data Mining, Baltimore, MD (November 2002)
[19] Elder, J., Pregibon, D.: A statistical perspective on kdd. In: Advances in Knowledge Discovery and Data Mining (1996)
[20] Eskin, E.: Anomaly detection over noisy data using learned probability distributions. In: Proc. 17th International Conf. on Machine Learning, pp. 255–262. Morgan Kaufmann, San Francisco (2000)
[21] Fayyad, U.M.: Mining databases: Towards algorithms for knowledge discovery. Bulletin of the IEEE Computer Society Technical Committee on Data Engineering 22(1), 39–48 (1998)
[22] Fayyad, U.M., Irani, K.B.: Multi-interval discretization of continuous attribous as preprocessing for classification learning. In: Proc. 13th Internat. Joint Conf. on Artificial Intelligence, Los Altos, CA, pp. 1022–1027 (1993)

[23] Friedman, N., Goldszmidt, M.: Building classifiers using bayesian networks. In: AAAI/IAAI, vol. 2, pp. 1277–1284 (1996)
[24] Fujikawa, Y., Ho, T.: Cluster-based algorithms for dealing with missing values. In: Chen, M.-S., Yu, P.S., Liu, B. (eds.) PAKDD 2002. LNCS, vol. 2336, pp. 535–548. Springer, Heidelberg (2002)
[25] Gary, K., Honaker, J., Joseph, A., Scheve, K.: Listwise deletion is evil: What to do about missing data in political science (2000), `http://GKing.Harvard.edu`
[26] Grzymala-Busse, J.W.: Rough set strategies to data with missing attribute values. In: Proceedings of the Workshop on Foundations and New Directions in Data Mining, the third IEEE International Conference on Data Mining, Melbourne, FL, November 2003, pp. 56–63 (2003)
[27] Grzymala-Busse, J.W.: Data with missing attribute values: Generalization of indiscernibility relation and rule induction. Transactions on Rough Sets 1, 78–95 (2004)
[28] Han, J., Kamber, M.: Data Mining Concepts and Techniques. Morgan Kaufmann, San Francisco (2006)
[29] Harms, S., Deogun, J., Saquer, J., Tadesse, T.: Discovering representative episodal association rules from event sequences using frequent closed episode sets and event constraints. In: Proceedings of the 2001 IEEE International Conference on Data Mining, San Jose, California, USA, November 29 - December 2, pp. 603–606 (2001)
[30] Hartigan, J., Wong, M.: Algorithm AS136: A k-means clustering algorithm. Applied Statistics 28, 100–108 (1979)
[31] Ho, L.S., Rajapakse, J.C., Nguyen, M.N.: Augmenting hmm with neural network for finding gene structure. In: Proc. of the 7th International Conference on Control, Automation, Robotics and Vision (ICARCV 2002), Singapore, pp. 1522–1527 (December 2002)
[32] Hullermeier, E.: Mining implication-based fuzzy association rules in databases. In: Proceedings of the 9th International Conference on Information Processing and Management of Uncertainty in Knowledge-based Systems, pp. 101–108 (2002)
[33] Ishibuchi, H., Yamamoto, T., Nakashima, T.: Fuzzy data mining: effect of fuzzy discretization. In: Proceedings IEEE International Conference on Data Mining, pp. 241–248 (November 2001)
[34] Jones, A.K., Sielken, R.S.: Computer system intrusion detection: A survey. Technical report, University of Virginia Computer Science Department (1999)
[35] Joshi, A., Krishnapuram, R.: Robust fuzzy clustering methods to support web mining. In: Proc. Workshop in Data Mining and knowledge Discovery, SIGMOD, pp. 15–1 – 15–8 (1998)
[36] Klawonn, F., Keller, A.: Fuzzy clustering based on modified distance measures. In: Hand, D.J., Kok, J.N., Berthold, M.R. (eds.) IDA 1999. LNCS, vol. 1642, pp. 291–299. Springer, Heidelberg (1999)
[37] Krishnapuram, R., Joshi, A., Nasraoui, O., Yi, L.: Low-complexity fuzzy relational clustering algorithms for web mining. IEEE Transactions on Fuzzy Systems 9(4), 595–607 (2001)
[38] Krishnapuram, R., Joshi, A., Nasraoui, O., Yi, L.: Low-complexity fuzzy relational clustering algorithms for web mining. IEEE Transactions on Fuzzy Systems 9(4), 595–607 (2001)

[39] Kumar, P., Krishna, P.R., Bapi, R.S., Kumar, S.: Rough clustering of sequential data. Data & Knowledge Engineering 63(2), 183–199 (2007)
[40] Kuok, C.M., Fu, A.W.-C., Wong, M.H.: Mining fuzzy association rules in databases. SIGMOD Record 27(1), 41–46 (1998)
[41] Li, D., Deogun, J., Spaulding, W., Shuart, B.: Dealing with missing data: Algorithms based on fuzzy sets and rough sets theories. Transactions on Rough Sets IV, 37–57 (2005)
[42] Li, D., Deogun, J., Wang, K.: Fads: A fuzzy anomaly detection system. In: Wang, G.-Y., Peters, J.F., Skowron, A., Yao, Y. (eds.) RSKT 2006. LNCS, vol. 4062, pp. 792–798. Springer, Heidelberg (2006)
[43] Li, D., Deogun, J., Wang, K.: Gene function classification using fuzzy k-nearest neighbor approach. In: Proceedings of the 2007 IEEE International Conference on Granular Computing (GrC 2007), San Jose, CA, pp. 644–647 (November 2007)
[44] Li, H., Zhang, W., Xu, P., Wang, H.: Rought set attribute reduction in decision systems. In: Wang, G.-Y., Peters, J.F., Skowron, A., Yao, Y. (eds.) RSKT 2006. LNCS, vol. 4062, pp. 135–140. Springer, Heidelberg (2006)
[45] Lingras, P., Yan, R., West, C.: Comparison of conventional and rough k-means clustering. In: Proc. of the 9th Intl Conf. on Rough Sets, Fuzzy Sets, Data Mining, and Granular Computing, Chongqing, China, pp. 130–137 (2003)
[46] Lippmann, R., Fried, D., Graf, I., Haines, J., Kendall, K., McClung, D., Weber, D., Webster, S., Wyschogrod, D., Cunningham, R., Zissman, M.: Evaluating intrusion detection systems: The 1998 DARPA off-line intrusion detection evaluation. In: Proceedings of the DARPA Information Survivability Conference and Exposition. IEEE Computer Society Press, Los Alamitos (2000)
[47] Little, R.J., Rubin, D.B.: Statistical Analysis with Missing Data. Wiley, New York (1987)
[48] Luo, J., Bridges, S.: Mining fuzzy association rules and fuzzy frequency episodes for intrusion detection. Intl. Journal of Intelligent Systems 15, 687–703 (2000)
[49] Matheus, C.J., Chan, P.K., Piatetsky-Shapiro, G.: Systems for knowledge discovery in databases. IEEE Trans. On Knowledge And Data Engineering 5, 903–913 (1993)
[50] Mitra, S., Pal, S.K., Mitra, P.: Data mining in soft computing framework: A survey. IEEE Transaction on Neural Networks 13(1), 3–14 (2002)
[51] Myrtveit, I., Stensrud, E., Olsson, U.H.: Analyzing data sets with missing data: an empirical evaluation of imputation methods and likelihood-based methods. IEEE Transactions on Software Engineering 27(11), 999–1013 (2001)
[52] Pawlak, Z.: Rough sets. International Journal of Computer and Information Sciences 11, 341–356 (1982)
[53] Perera, A., Denton, A., Kotala, P., Jockheck, W., Granda, W., Perrizo, W.: P-tree classification of yeast gene deletion data. SIGKDD Explorations (2002)
[54] Portnoy, L., Eskin, E., Stolfo, S.: Intrusion detection with unlabeled data using clustering. In: ACM Workshop on Data Mining Applied to Security (2001)
[55] Roth, P.: Missing data: A conceptual review for applied psychologists. Personnel Psychology 47(3), 537–560 (1994)
[56] Schafer, J.L.: Analysis of Incomplete Multivariate Data. Chapman & Hall/CRC, Boca Raton (1997)

[57] Shahbaba, B., Radford, M.N.: Gene function classification using bayesian models with hierarchy-based priors. Technical Report 0606, Department of Statistics, University of Toronto (May 2006)

[58] Sim, J., Kim, S.-Y., Lee, J.: Prediction of protein solvent accessibility using fuzzy k-nearest neighbor method. Bioinformatics (21), 2844–2849 (2005)

[59] Slowinski, R., Vanderpooten, D.: A generalized definition of rough approximations based on similarity. IEEE Transactions on Knowledge and Data Engineering 12(2), 331–336 (2000)

[60] Störr, H.-P.: A compact fuzzy extension of the naive bayesian classification algorithm. In: Proc. In Tech/VJFuzzy 2002, Hanoi, Vietnam, pp. 172–177 (2002)

[61] Vinayagam, A., Konig, R., Moormann, J., Schubert, F., Eils, R., Glatting, K.H., Suhai, S.: Applying support vector machines for gene ontology based gene function prediction. BMC Bioinformatics (5) (2004)

[62] Weiss, S.M., Indurkhya, N.: Decision-rule solutions for data mining with missing values. In: IBERAMIA-SBIA, pp. 1–10 (2000)

[63] Yager, R.R.: Using fuzzy methods to model nearest neighbor rules. IEEE Transactions on Systems, Man and Cybernetics, Part B 32(4), 512–525 (2002)

[64] Zadeh, L.A.: Fuzzy sets. Information and Control 8, 338–353 (1965)

[65] Zeng, H., Lan, H., Zeng, X.: Redundant data processing based on rough-fuzzy. In: Wang, G.-Y., Peters, J.F., Skowron, A., Yao, Y. (eds.) RSKT 2006. LNCS, vol. 4062, pp. 156–161. Springer, Heidelberg (2006)

[66] Ziarko, W.: The discovery, analysis and representation of data dependencies in databases. In: Knowledge Discovery in Databases, pp. 195–209. AAAI Press, Menlo Park (2000)

Data Driven Users' Modeling

Danuta Zakrzewska

Abstract. Together with the development of Internet systems and services, there appeared a necessity of designing Web applications that fulfil users' information needs. These requirements are specially important in case of educational systems and portals, fulfilling them may be obtained by building users' models and personalizing systems. Learners may be modeled on the basis of data, contained in log files, registered during their historical activities, or on the basis of data collected as questionnaires' results, that in case of educational systems may be connected with learning styles preferences. In both approaches, data driven models allow for grouping users and differentiating computer programs to satisfy their requirements. Application of data mining techniques enables to find behavioral patterns of users as well as indicating groups with similar features.

In the chapter, modeling on the basis of both kinds of data sources is considered. Log data are used for finding the most frequent users' navigational paths by application of frequent pattern mining. There is proposed improvement of that technique by using OLAP operations to preprocess data. Application of cluster analysis for finding groups of users with similar navigational patterns, is presented. Performance of clustering algorithms is also investigated for learners modeling according to their learning styles data.

1 Introduction

Development of Web services resulted in building systems for the wide range of users, with different information needs. Growth of e-commerce services as well as the big competition in the business area, directed the attention of application

Danuta Zakrzewska
Institute of Computer Science Technical University of Lodz, Wolczanska 215, 90-924 Lodz, Poland
e-mail: dzakrz@ics.p.lodz.pl

D. Zakrzewska et al. (Eds.): Meth. and Support. Tech. for Data Analys., SCI 225, pp. 115–136.
springerlink.com

designers into the personalisation features of computer programs. Together with e-business systems' development, there appeared also many virtual communities centered around Internet portals, and learning communities constitute one of the most numerous group among them. At the beginning, portals attract the members, but their sustainability becomes a big problem and acceptance of technology may play a significant role in it. In [46], Teo, Chan, Wei & Zhang introduced extended TAM (Technology Acceptance Model) of sustainability of virtual learning communities. They indicated that information content, the type of communication channels provided and information organization are important success factors. User centered adaptation ability of a portal, which takes into account all of these factors, may help to achieve information accessibility satisfying users' expectations. Building profiles and segmenting users according to their preferences enable tailoring the content and information organization of portals in the way that may fulfil users' needs and requirements.

In many cases, virtual learning communities are connected with distance education. In that area, the main requirement for educational software is to achieve the pedagogical goal. But except of fulfilling pedagogical needs, portals should have ability of customization according to individual users' preferences, as personalization holds great potential to improve people's ability to interact with information [24].

The research in the area of building personalised learning environments is conducted in two main directions. The first one focuses on determining individual usability preferences, while the second one concerns modeling students according to their individual characteristics, which may be found out by using different kinds of questionnaires, mostly focused on investigating users' cognitive capabilities.

Usability preferences are usually determined by application of web mining methods. Broad range of articles describing these investigations may be found in [38]. Research, in that area, are very often based on data collected during historical users' activities and contained in log files. In last years, these files have become the main data sources for examinations of users' preferences. Many authors investigated techniques for categorization of users according to their information needs, by extracting knowledge from logging data (see [31],[35],[42],[51]). As the main goals of the research in this area, there should be mention building adaptive Web sites or Web recommender systems. The investigations concerning mining for identification of users' needs were also conducted for information systems designing (see [28]) as well as workflow process building (see [1] for example). According to the research experience so far, logs occurred to be the good data sources for mining user preferences in all considered cases. There are two main directions in log data mining: finding navigational patterns and grouping users according to their preferences. There exist different approaches for analyzing log data, as main techniques there should be mentioned association rules, sequential patterns and clustering. The first two groups of methods are usually applied for finding navigational patterns, while cluster analysis is used for data segmentation and for finding groups of users with similar preferences(see [8]). The wide review of all the methods and recent development in the area, are presented in [9],[10],[33] and [36].

There exist different factors that may influence learners' individual attitudes towards using the software and taking active part in educational process, as well as obtaining assumed learning outcomes. Cognitive features, such as personal learning styles may indicate the directions in differentiating of educational proccess, and adaption of teaching materials into individual needs may be done according to them. An overview of the research concerning building adaptive systems by using learning styles can be found in [43]. The most popular method, consists in assigning students into predefined groups, without possibility of updating (compare [14],[40],[47]). Application of data mining techniques allows for creating students' groups according to their individual characteristics, what in turn, enables preparing customised educational materials and constructing individual teaching paths.

In the chapter two approaches to users' modeling are presented: finding groups of users with similar navigational patterns on the basis of their historical activities and grouping learners according to their learning styles preferences. In the first case, the research is done for data registered in log files, while in the second one data collected as questionnaires' results are investigated. Paths for both of the approaches are presented on Fig. 1. In each of the considered cases data mining algorithms are proposed.

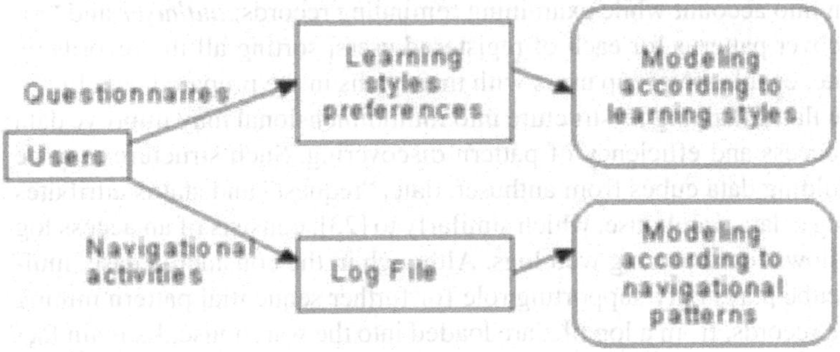
Questionnaires
Learning styles preferences
Modeling according to learning styles
Users
Navigational activities
Log File
Modeling according to navigational patterns

Fig. 1 Two paths of users' modeling

2 Modeling according to Navigational Patterns

Data, used for users' profiling according to their navigational paths are usually collected *implicitly* by recording the navigation requests in log files and in that case obtained models are rather dynamic.

2.1 Log Files

Each access to a Web page is recorded on the server that hosts it in the log file([9]). We consider log data, written on the Web server, in Common Logfile Format with records in the form ([30]) :

remotehost rfc931 authuser date "request" status bytes,

where

remotehost is remote hostname (or IP number if hostname not available),
rfc931 is remote logname of the user,
authuser is the username as which the user has authenticated himself,
date is date and time of the request,
"request" is the request line exactly as it came from the client,
status is the HTTP status code returned to the client,
bytes the content-length of the document transferred.

For further considerations, which main goal is to find portal participants' navigational models, data should be cleaned of the useless attributes and faulty records. From among the first three fields, only the third one signifies authorized name of the portal participants. The other two are connected with remote computers and may not characterize users properly. The value of attribute status, decides of the correctness of the data, equal to 200, indicates that the client's request was successfully received, understood, and accepted, therefore only these records should be taken into account. The next filter should be put on the date attribute. However all requests are registered succesively and time order is maintained, but filters are neccessary for restricting considered time intervals. In final investigations, there are only two attributes taken into account while examining reminding records: *authuser* and "*request*". To discover patterns for each of registered users, sorting all the records by the first attribute, enables to group users with their paths in the prepared data file.

Log files are flat. Changing its structure into multidimensional may improve data preparation proccess and efficiency of pattern discovering. Such structure may be obtained by building data cubes from authuser, date, "request" and status attributes and constructing a data warehouse, which similarly to [23], consists of an access log part and that allows for analysing web logs. Although in the considered case, multidimensional cube plays only supporting role for further sequential pattern mining operations. The records, from a log file, are loaded into the warehouse. Its main fact table contains keys to related dimensions: authuser, date ,status and url of "request". The view of the multidimensional star schema is presented on Fig. 2.

In the multidimensional data cube, the attribute date, of a log record, was changed into the time dimension, with the hierarchy of month, day and hour extracted from

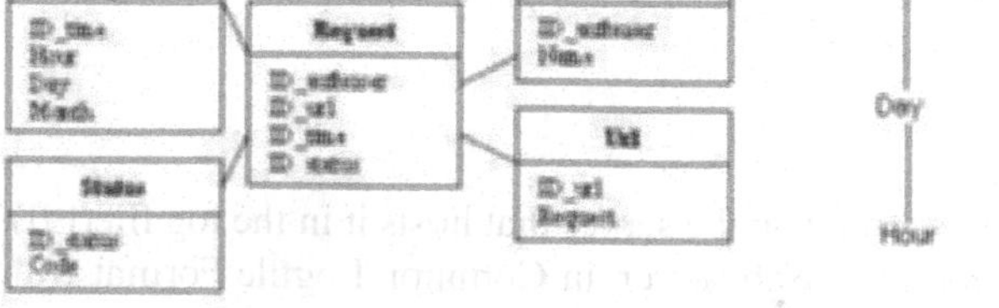
Time hierarchy
Request
Month
Day
Hour

Fig. 2 Multidimensional schema for request facts and hierarchy of time dimension

the date fields (see Fig. 2). Such defined hierarchy may help in discovering patterns for required time periods or day hours.

2.2 Finding Navigational Patterns

The choice of technique for finding users' behavioral patterns depends mostly on the goal of the research. Cluster analysis is usually used for grouping data according to similar sessions or groups of sessions, which are represented by sets of visited Web sites (see [13],[49],[50]). In that approach, the main difficulties, consist, not only, in the right choice of similarity functions and features, which characterize sessions, but also in the fact that different users may belong to more than one cluster [34], what is difficult to achieve for commonly used clustering methods. What is more, clustering algorithms need special data preparation, they depend on order of input data and they are computationally expensive [36]. Usage of association rules for discovering navigational patterns is limited, because that technique does not allow for taking into account time order of users' requests. The requirement concerning inclusion of time elements into discovered patterns may be fully realized, by application of sequential pattern mining techniques, that identify frequent sequences of events, which are visits on web sites. There exist different approaches in this area: as the most frequently used, it should be mentioned algorithms based on association rules, methods using tree structures and Markov chains. The last techniques can generate navigation paths that may be used automatically for prediction, but they do not produce readable user models and are computationally expensive (see [36]). Tree structures are usually applied in recommender systems, like WAP-tree [35] or NP-tree [21], which enable discovering behavioral patterns in a real time. Sequential pattern mining techniques based on association rules mainly use *Apriori* algorithm [16] and its modifications. However they are not efficient in case of long patterns [35], but they are scalable and result in models, which are easy to read, what is of some significance in further users' profiles building. Most of authors focus their research concerning users' navigational patterns on application of one methodology, however some of personalization systems use different methods, but separately (see [31]). In this study, we will focus on the approach, which consists on combining two different techniques: sequential pattern mining and cluster analysis. Such connection of the methods allows for building navigational patterns and for discovering groups with similar interests at the same time. Sequential pattern mining contrarily to association rules, which do not take into account the time sequence, allows for using events in the order they happened and seems to be more convenient for finding such patterns as users navigation paths. What is more, this method may help in discovering all the frequent subsequences from all sequences of pages visited by users (compare [42],[51]), while clustering algorithm, applied to navigational patterns, allows for segmenting almost all of the portal participants and organizing the information content , in the way that fulfils their needs.

2.3 Sequential Pattern Mining

To find out navigational preferences of users, there will be used the algorithm adopted from sequential pattern mining technique, introduced in [2] to investigate ordered sequences of items, which were purchased by particular clients. The applied method is based on resolving the issue of discovering sequential patterns into the problem of finding maximal sequences with the minimal support, from among all existing sequences of users.

Definition 2.1. *Let url_i means the i-th request of the user, placed in the field "request" of a log file records. The request is composed of the set of single documents $x_j(url_i), j = 1, 2, ...,$. Let us denote it as $X_i = X(url_i)$. The ordered list of sets $X_i, i = 1, 2, ...,$ will be called the sequence S. Let D be the set of all sequences contained in the log file.*

Definition 2.2. *Let for fix $j, j = 1, 2, ...; S_j$ be a sequence from D. We will say that a sequence $S \in D$ supports S_j, if S_j is contained in S, $S_j \subset S$. S_j will be called subsequence of S. Number of all subsequences of S belonging to D will be called a support of the sequence S.*

Definition 2.3. *Number of sets composing the list of the sequence S is called its length j. A sequence of lenght j is called j-sequence.*

Definition 2.4. *We will say that a sequence S is maximal, if it is not contained in any other sequence. S is called a frequent pattern if it is maximal and its support value is not less than required minimal number.*

Sequences and frequent patterns for data contained in log files, determined by the above definitions, enable using introduced in [2] technique, to find out frequent patterns as maximal sequences with support values greater or equal to required minimum. In our approach, each sequence of *url*, consists of subsequences of requested documents. Users' navigational patterns are the sequential patterns, which are represented by maximal sequences from among all the sequences with a certain minimal support value, as it is specified in Def. 2. We look for preferred navigational paths, determined by maximal sequences, in the set of all the sequences of each user. The algorithm may be presented as follows [52]:

For each user:
Input:
S - the set of all user's sequences
maxlength - the length of the longest sequence.
minsup - minimal required support value
Steps:
$S0 = S$
For$(k = maxlength; k > 1; k--)$ ***do***
Foreach *k-sequence* s_k ***do***

{
***Find** all subsequences s_k in S*
***Remove** all subsequences s_k from S.*
Support(s_k) = count(subsequences s_k in S0)
***if** Support (s_k) $\geq$ minsup **then** indicate s_k as a pattern*
}
Output:
Patterns and their supports

As the results of the algorithm, we obtain maximal sequential patterns and their supports for each registered user. New database with records of three attributes: *authuser*, its pattern and its support, is created. It contains much less number of records than there are contained in log files. Attributes represent navigational preferences of each user and can be used for building users' profiles by further application of filter operations or by clustering similar users.

2.4 Application of OLAP Operations

Application of sequential pattern mining on flat log files may be not effective in case of their big sizes. Building a multidimensional data cube and using OLAP operations as preprocessing to data mining technique can definitely improve the process [53]. The similar approach was described in [51]. The system WebLogMiner, presented there, enables building multidimensional cubes from web log data for further using of data mining methods, but the tool, was designed only for a custom multidimensional database. Connecting of online analytical processing with web usage mining was also examined by Büchner and Mulvenna [5] for e-commerce purposes. In [23], a proposed data warehouse was enhanced in an *ad-hoc* tool of analytic queries for web logs analysing. Multidimensional structure, presented there, was used to generate association rules for web pages and for clustering of users' sessions.

In our approach the multidimensional structure (see Fig. 3) presented in Sec. 2.1 allows for using OLAP operations for the purpose of preparing data into sequential pattern mining techniques. The first of considered OLAP functions is slicing operation used to *Status* dimension and its goal is to filter all the "requests" that

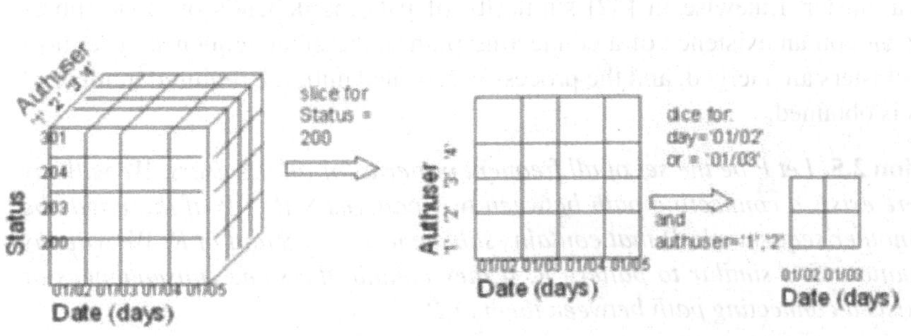
Authuser
'1' '2' '3' '4'
Status
301
204
203
200
01/02 01/03 01/04 01/05
Date (days)
slice for
Status =
200
Authuser
'1' '2' '3' '4'
01/02 01/03 01/04 01/05
Date (days)
dice for
day='01/02'
or = '01/03'
and
authuser='1','2'
01/02 01/03
Date (days)

Fig. 3 Slicing and dicing for exemplary log data

ended without any error, what takes place in case of *HTTP Status code = 200*. The next OLAP operation used for sliced data was dicing applied to *Authuser* and *Time* dimensions, it enables finding preferences of the certain users or groups of users depending on months, days or hours. The usage of the both of OLAP operations is presented on Fig. 3.

Using of OLAP operations prepares data for effective application of the algorithm introduced in Sec. 2.3, without any additional preprocessing.

2.5 Clustering

Application of cluster analysis for portal participants allows for segmenting them according to their information needs and preferences. Most of the papers concerning finding groups of web users are based on clustering of web sesions (see for example [13],[49],[50]). Paliouras et al., in [34], additionally to users' segmentation, identified the distinguishing characteristics of each discovered cluster. They compared three different cluster analysis algorithms for constructing communities with similar preferences. The main problem in clustering web users consists in the fact that the same user may belong to two groups of different preferences and, then, should be assigned to two different clusters. In the presented research, users are clustered according to their sequential patterns of navigational paths. The similar approach was presented in [7], but paritionning clustering technique was based on *Frequently Based Subtree Patterns (FCSPs)* which represent characteristics in the evolution of the usage data. In our considerations, we will present grouping users by clustering their navigational patterns by using a hierarchical agglomerative technique, based on the *Pattern Oriented Partial Clustering* algorithm introduced in [32]. That way, we combine two different techniques, what allows for building navigational patterns and for discovering the groups with similar information preferences at the same time. Finding sequential patterns helps in discovering all the frequent subsequences from all sequences of pages visited by users. Clustering algorithm, in turn, allows for segmenting almost all of the portal participants and for organizing the information content that will fulfill their needs.

All frequent patterns found during sequential pattern mining, applied for all users, are grouped by using hierarchical agglomerative algorithm, starting from each pattern as a cluster. Likewise in [32] similarity of patterns depends on their subsequences and on an existence of a connecting path in the other sequence. The most similar clusters are merged, and the process is repeated until the required number of clusters is obtained.

Definition 2.5. *Let P be the set of all frequent patterns for all the users. We will say that there exists a connecting path between two patterns $S, R \subset P$, if there may be found another sequence in P that contains subsequences of S and of R. We will say that a pattern S is similar to pattern R, if they contain the same subsequences or there exists a connecting path between them in P.*

Definition 2.6. *Let c_i and c_j be two clusters containing frequent patterns Ṡimilarity between c_i and c_j is defined by Jaccard's coefficients:*

$$J(c_i, c_j) = \frac{|c_i \cap c_j|}{|c_i \cup c_j|}. \tag{1}$$

where $c_i \cap c_j$ denotes the intersection of clusters, its size represents the number of patterns that are contained in both clusters and are similar to each other; $c_i \cup c_j$ denotes the union of clusters, its size represents number of patterns from both clusters, except those, who are similar.

Clustering algorithm may be presented as follows:

Input:
P - the set of all users' frequent patterns
NP - number of frequent patterns
N - number of required clusters
Steps:
Assign each pattern to a separate cluster. Denote by C_1 the initial set of clusters
$k = 1$
while $|C_k| > N$ ***do***
foreach $(c_i, c_j) \in C_k$ ***do***
{
count $J(c_i, c_j)$
find i_{max}, j_{max} *such that* $J(c_{i_{max}}, c_{j_{max}}) = max\{J(c_i, c_j) : c_i, c_j \in C_k\}$
merge clusters $(c_{i_{max}}, c_{j_{max}})$
add the new cluster to C_k
remove $(c_{i_{max}}, c_{j_{max}})$ *from* C_k
$C_k = C_{k+1}; k{+}{+}$
}
Output:
Set of clusters C_N

Similarity values do not have to be counted in every stage during a process of building a new set of clusters. Using of similarity matrix, which stores all the Jaccard's coefficients may allow for omitting that step for every two clusters that were not taking part in a merging process [32]. Similarity matrix is a square symmetrical matrix. At the beginning, its dimension is equal to the input number of frequent patterns and, during every iteration, is reduced until N in the output. Creating of the similarity matrix requires calculating of $(NP^2 - NP)/2$ Jaccard's values, in the first step, while in every next iteration only counting of $(NP - 1)$ values is neccessary.

Obtained clusters contain similar navigational paths in the form of sequences, together with users connected to them. Analysis of these groups allows for designing different information contents of portals for users with similar preferences. However users are usually connected to more than one sequential pattern and it may happen that they are allocated into more than one cluster, or they may not be assigned into any group of preferences at all.

All the stages of the described technique of finding users' profiles, starting from a log file, are presented on Fig. 4.

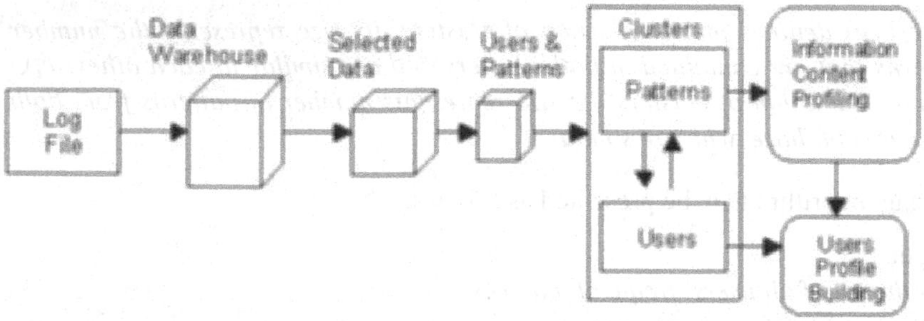
Data
Warehouse
Selected
Data
Users &
Patterns
Clusters
Log
File
Patterns
Users
Information
Content
Profiling
Users
Profile
Building

Fig. 4 Stages of the process of finding users' profiles

2.6 Evaluation

Evaluation of the performance and the efficiency of the presented method was done, on the basis of experiments conducted for the files containing records of authorised login registrations to the portal: `http://machines.hyperreal.org`, during the period of 4 months. Previously, these log files were used in the investigations concerning adaptive web sites and they are available at `http://www.cs.washington.edu/ai/adaptive-data` [56]. The technique was examined by using files of different sizes: 1MB, 2MB, 3MB, 4MB and 10MB, as well as specially prepared samples, what allowed for the analysis of obtained patterns and clusters. To preserve the privacy, all the records have been anonymized, by converting authorised users' names to meaningless numbers. All the experiments were conducted on the personal computer equipped in processor Intel Celeron 1.47GHz and 448MB RAM, with operating system Windows XP. The method was examined by reviewing each of the considered steps: sequential pattern mining, preprocessing data by OLAP and clustering.

Experiments connected with the first phase consisted in investigating the run time of sequential pattern mining as well as the number of obtained patterns as functions of log file sizes. The left graph presented on Fig. 5 shows that the run time of finding sequential patterns depends almost linearly on log file sizes. This feature is connected with the increasing number of sequential patterns together with the growth of file sizes, as is presented on the right-hand side of Fig. 5. Graphs on Fig. 5 do not take into account support values as they do not influence the run time of sequential pattern mining step. What is more, examinations of the results indicated that support values for different patterns do not depend on the file sizes, but on their structures. Estimation of the quality of obtained results for real data, in case of a big amount of users and their frequent visits of web sites, is difficult for realizing. The tests were

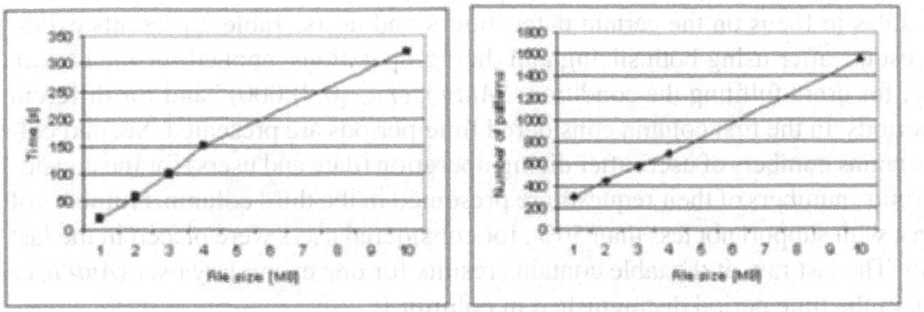

Fig. 5 Runtime and number of the sequential patterns depending on the log file size

done on specially prepared log files, with small number of users and their paths. The results were consistent with expected, however the dependance of the number of patterns and their support values on the number of registered paths was observed. For example, in case of small number of records the longest users' paths have usually bigger support values and more chance to be recognized as patterns. Analysis of logs' structures together with obtained results showed that only the big amount of users' requests may give reliable patterns, and such requirement means big log file sizes.

The main aim of the second step of experiments, was to investigate the influence of building multidimensional data cube and application of OLAP operations, in the pre-processing phase, on the performance of the sequential pattern mining algorithm. Application of multidimensional data structure enables focusing on certain users or groups of users and on their behaviours in determined periods of time or hours during the day, what may be of some significance in users' profiles building. That phase occured to be important, when taking into account execution time of the sequential pattern mining algorithm, that grows linearly together with file sizes (see Fig. 5).

Table 1 presents exemplary results after application of a slicing operation with the criterion: "*HTTP Status code=200*". It can be easily seen that numbers of records decreased from 24% to 32%, depending on the file size. Usage of the dicing operation on sliced data allows for further diminishing of number of considered records

Table 1 Number of records for different file sizes after slicing operation

No	File size	Number of records	Records with status=200	New file size
1	1MB	4761	3241	0.7MB
2	2MB	9383	6341	1.4MB
3	3MB	13200	9915	2.3MB
4	4MB	17506	13350	3.1MB
5	10MB	44850	32850	7.3MB

and enables to focus on the certain dates, hours and users. Table 2 presents exemplary results, after using both slicing and dicing operations, applied for the certain log file, for users fulfiling the condition: "*Authuser* $\in \langle 0, 10000 \rangle$" and for different time periods. In the first column considered time periods are presented. Second column contains numbers of users after dicing operation (date and users) for the considered group , numbers of their requests are presented in the third column. Numbers of patterns, with support not less than 50%, for considered users were placed in the last column. The last raw of the table contains results for one exemplary user (*Authuser* = 105), in the time period distinguished in column 1.

Table 2 Users and patterns after slicing and dicing operation

Time period	Number of users	Users requests	Patterns
1999.02.02	281	2366	204
1999.02.02,04.08,04.10(16:00:00..23:59:59)	329	2976	248
1999.02.02-(16:00:00 .. 23:59:59)	184	1365	134
1999.02.02..1999.04.10	1	47	1

The experiments showed that using of OLAP operations significantly shortened the run time of the sequential mining algorithm, especially dicing operation applied for attributes: "*date*" and "*Authuser*" occured to be very effective. However, limiting only to certain time periods, by using slicing operation to attribute "*date*", enables decreasing number of considered records, but examining the requests only of users from determined groups allows for significant diminishing of investigated data. Comparison of number of requests for different dates presented in Table 3 with column 3 of the Table 2 shows that in the examined case number of records decreased more than 80% after limiting into determined groups of users. Additionally using of dicing operation allows for focusing on the certain groups of users navigating in defined time periods, what may help in further analysis and application of obtained results.

Table 3 Requests after slicing according to "*date*"

Date	Number of requests
1999.02.02	12902
1999.02.02,04.08,04.10(16:00:00..23:59:59)	17423
1999.02.02-(16:00:00 .. 23:59:59)	8091

In the last step of experiments, execution time of the clustering as well as the quality of obtained clusters were considered. Research was done for different log file sizes and different required number of clusters. Investigations showed that the run time of the clustering stage increases together with the growth of log file sizes (see Fig. 6), what is connected with the increasing number of sequential patterns. Thus, application of OLAP in the preprocessing stage may play significant role in diminishing clustering run time in case of big file sizes.

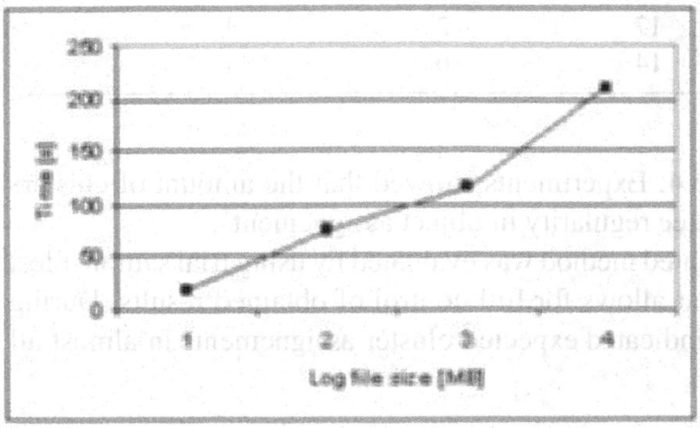
250
200
150
100
50
0
1
2
3
4
Log file size [MB]

Fig. 6 Clustering time for different log file sizes and number of clusters=5

Investigations showed that the dependence of the execution time on the required number of clusters occured to be small, for files of not big sizes. For example, in case of 1MB volume, run time differences become observable, if the number of clusters are equal, at least to 100 or more. The differences are growing together with file sizes, what is presented on Fig. 7.

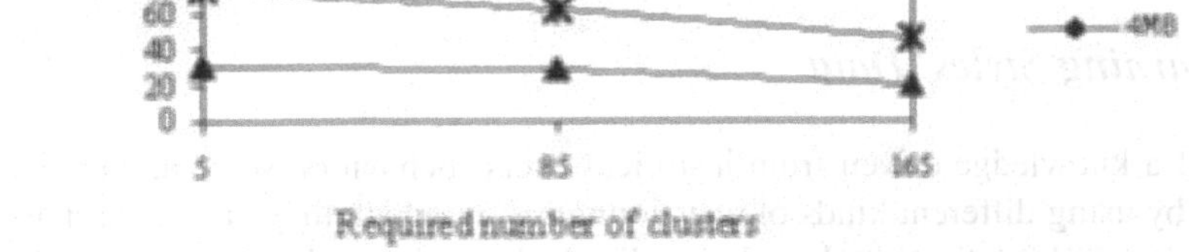
240
220
200
180
160
140
Time [s] 120
100
80
60
40
20
0
5
85
165
Required number of clusters

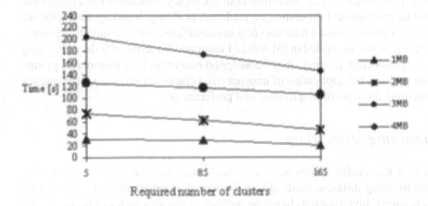

Fig. 7 Clustering time for different log file sizes and number of clusters

Examinations of the quality of assignements, depending on the required number of clusters, indicated that in case of their big amount, obtained groups are of different sizes. Almost all the users are allocated into the biggest cluster, while all the others contain only few objects. Exemplary amount of clusters containing different number of objects for required number of clusters equal to 25 and for different log

Table 4 Amount of clusters of different volumes. Required number of clusters=25

Log file size	1-2 objects	3-5 objects	6-10 objects	> 10 objects
1MB	9	14	1	1
2MB	8	11	5	1
3MB	5	17	2	1
4MB	4	14	6	1

file sizes is presented in Table 4. Experiments showed that the amount of clusters equal to 10 or less may guarantee regularity in object assignements.

The correctness of the presented method was evaluated by using trial sample files, containing several records what allows for full control of obtained results. During tests the clustering algorithm indicated expected cluster assignements in almost all of considered cases.

3 Modeling according to Learning Styles

Building students' profiles according to their preferred learning styles' dimensions enables individualization of teaching process, by constructing personal educational materials and by designing individual learning paths, what is of special significance, in case of distance education. Impact of learning styles on web based courses' performance, was identified by many authors (see [27],[29]). Beaudoin & Dwyer [3] stated that a necessary element for online course efficacy is establishing a collaborative learning environment depending on students' profiling. Building an intelligent learning environment should take into account interface customization to learners' preferences as it was described in [6], where Felder and Silverman model of learning styles was considered. Learners may be assigned into artificially formulated groups of certain profiles, but application of unsupervised classification of students enables for segmenting them according to their real preferences.

3.1 Learning Styles' Data

In spite of a knowledge driven from historical users' behaviors, students may be modelled by using different kinds of questionnaires, in which they answer various sets of questions. Investigations based on individual learning styles of users are one of the most often used. The relationship of individual learning styles and distance education was examined by different researchers, who indicated learning styles as valuable features of users' models in a teaching process (see for example [11]). Graf & Kinshuk [15] showed that students with different learning styles have different needs and preferences, and as was stated by Rovai [39] teaching methods should vary according to them.

There exist different models of learning styles such as Felder & Silverman [12], Kolb [26], and Honey & Mumford [20], to mention the most frequently used. In our investigations, Felder and Silverman model, which is often used for providing adaptivity regarding learning styles in e-learning environments [48], is presented. It is based on *Index of Learning Style* (ILS) questionnaire, developed by Soloman & Felder [41]. The results of ILS questionnaire indicate preferences for four dimensions of the Felder & Silverman model: *active* vs *reflective*, *sensing* vs *intuitive*, *visual* vs *verbal*, *sequential* vs *global*. The index obtained by each student has the form of the odd integer from the interval [-11,11], assigned for all of the four dimensions. Each student, who filled ILS questionnaire, can be described by a vector SL of 4 integer attributes:

$$SL = (sl_1, sl_2, sl_3, sl_4) = (l_{ar}, l_{si}, l_{vv}, l_{sg}). \tag{2}$$

where l_{ar} means scoring for *active*(if it has negative value) or *reflective*(if it is positive) leraning style, and respectively l_{si}, l_{vv}, l_{sg} are points for all the other dimensions, with negative values in cases of *sensing*,*intuitive* or *visual* learning styles, and positive values in cases of *intuitive*,*verbal* or *global* learning styles.

Thus, further considerations concerning clustering students with similar preferences will take into account data in the form described by (2).

3.2 Cluster Analysis

Students' models described by (2), are simple and the usage of one of well known cluster analysis techniques to find students groups seems to be the natural choice. However, examining different approaches to students' clustering, on the test data sets, showed big dependance of obtained results on the choice of input parameters as well as big noise sensitiveness of algorithms [55]. Especially the last feature of the algorithms may cause many problems if data contain noise or if outstanding students belong to the considered group. In case of partitioning methods as well as hierarchical techniques the wrong choice of required number of clusters may result in a low quality of obtained clusters. Density based approach, like DBSCAN (Density Based Spatial Clustering of Applications with Noise), may be very effective in case of large multidimensional data sets with noise and built clusters of arbitrary shapes [16], but strongly depends on the parameters connected with the density of objects, that are difficult to determine.

To avoid low quality of obtained clusters caused by the presence of noisy data, preprocessing phase, which consists of finding and indicating outliers and of removing them from the data set, is proposed.

3.3 Dealing with Outliers

In the literature there exist many approaches for detecting outliers, the most important of them are: distance based [25],[37][45], density based [4],[22] and cluster based [17],[44]. The broad review of modern methods of detecting outliers was presented in [19]. In the current study, the modification of distance based Connectivity Outlier Factor (COF) is proposed [54], because of its simplicity and big efficiency, in comparison to other approaches. COF algorithm was introduced in [45], as the result of an observation that although high density of data may represent a pattern, all patterns are not necessarily of high density. Density based methods may indicate as outliers objects, which are close to patterns of low density. The COF indicator determines how much an object is isolated from the others and thus in what degree it can be regarded as an outlier.

Definition 3.1. *Let O be the set of objects. Let $P, Q \subset O$ and $P, Q \neq \emptyset$ and $P \cap Q \neq \emptyset$. Distance between P and Q is defined as: $d(P,Q) = min\{d(x,y) : x \in P \wedge y \in Q\}$. Distance between $p \in P$ and Q is defined as:$d(p,Q) = min\{d(p,y) : y \in Q\}$.*

For any $q \in Q$ we say that q is a nearest neighbour of P in Q if there exists $p \in P$ such that: $d(p,q) = d(P,Q)$.

Definition 3.2. *Let $\{p_1, p_2, ...p_r\}$ be a subset of G. A set-based nearest path or SBN-path, from p_1 on G is a sequence $\{p_1, p_2, ...p_r\}$, that for all $i \in [1, r-1]$ element p_{i+1} is the nearest neighbour of the set $\{p_1, p_2, ...p_i\}$ in $\{p_{i+1}, p_{i+2}, ...p_r\}$. An SBN-path indicates the order in which the nearest objects are presented.*

Definition 3.3. *Let $s = \{p_1, p_2, ...p_r\}$ be an SBN-path. As a set based nearest trail or SBN-trail with respect to s we will consider a sequence of edges $\{e_1, ..., e_{r-1}\}$ such that for all $i \in [1, r-1], e_i = (o_i, p_{i+1})$ where $o_i \in \{p_1, p_2, ...p_i\}$ and $d(e_i) = d(o_i, p_{i+1}) = d(\{p_1, p_2, ...p_i\}, \{p_{i+1}, p_{i+2}, ...p_r\})$. We call each e_i an edge and the sequence $\{d(e_1), ..., d(e_{r-1})\}$ the cost description of $\{e_1, ..., e_{r-1}\}$.*

Definition 3.4. *Let $s = \{p_1, p_2, ...p_r\}$ be an SBN-path from p_1 and $e = \{e_1, ..., e_{r-1}\}$ be the SBN-trail with respect to s. The average chaining distance from p_1 to $G - \{p_1\}$ is defined as:*

$$ACD_G(p_1) = \sum_{i=1}^{r-1} \frac{2(r-i)}{r(r-i)} d(e_i). \tag{3}$$

Definition 3.5. *Let $p \in P$ and $k \in N$. The connectivity-based outlier factor COF at p with respect to its k-neighbourhood N_k is:*

$$COF_k(p) = \frac{|N_k(p)| \cdot ACD_{N_k(p)}(p)}{\sum_{o \in N_k(p)} ACD_{N_k(o)}(o)}. \tag{4}$$

According to Def. 3.5, COF indicates how far from a pattern an object is situated. It is shown in [45] that for objects belonging to the pattern COF value is close to 1. Objects with COF greater than 1, with high probability, may be classified as outliers. To determine outliers, a parameter called outlierness factor was introduced in [54].

Definition 3.6. *Let O be a set of n objects $o_i \in O$, and $o \in O$. Let $COF(o)$ denotes the COF value of o. Then $COF_{max} = max\{COF(o_1), COF(o_2), ..., COF(o_n)\}$. We will consider object o as an outlier if its local outlier factor (COF) is such that:*

$$COF(o) > COF_{max} - \frac{COF_{max} - 1}{OF}, \quad (5)$$

where OF is the value of outlierness factor.

Introducing OF enables indication of outliers, when their number is unknown. Despite of density based methods, COF is capable to detect outliers among data of different density, what is more, COF technique is able to indicate outliers situated close to high density area. Investigations described in [52] showed that distance based COF algorithm is effective for finding outliers in students' learning styles data. The technique may not only indicate noise, but also outstanding students that should be taught by using individual paths.

3.4 Experiments

To evaluate the efficiency of the presented approach, the experiments were done on the real students' data sets. Investigations, described in [55], showed that, if data do not contain outliers, partitioning methods as well as hierarchical ones presented good performance on test data sets. After eliminating outliers, in the second stage of students clustering according to their learning styles, Farthest First Traversal algorithm was applied. This hierarchical divisive approach was introduced in [18]. The algorithm selects the most diverse objects, by choosing the first cluster centre at random, the next centres are the points situated farthest away from the current set of selected objects, where the distance is measured as determined in Def. 3.1. The process is repeated until the required number of clusters K is obtained. If K is small and the number of classified objects is not very big this technique occurs to be effective.

During experiments choice of input parameters and the performance of the considered two stage technique were examined. Research was done on the training data set of 125 Computer Science students' learning styles. The examined input parameters that influence the system performance were connected with students segmentation process.

In the first stage, parameters which decide of the number of outliers indicated by COF algorithm were investigated, as they are crucial for the structure of groups obtained in the second stage. Two values decide on the correctness of the choice of outliers: number of neighbors k and outlierness factor OF. Research concerning credit cards' data presented in [54] has shown that growing value of k improves the effectiveness of the technique of detecting the proper number of outliers. In case of learning styles dimensions, experiments conducted for data with different degrees of noisiness demonstrated that performance of the method depends on proper choice of both values, k and OF, at the same time. The greater k is chosen, the bigger range of OF value may guarantee good results. For example, for the considered data set, if $k = 15$, OF should belong to the interval $\langle 3, 22 \rangle$, while for $k = 8$, only OF equal to 3 or 4, gives satisfying effects. Some exemplary data are presented in Table 5 and Table 6. In both of the tables, data are sorted according to COF values. It can be

Table 5 Elements of a vector *SL* and *COF* values for $k = 8$

sl_1	sl_2	sl_3	sl_4	*COF*
11	11	11	-11	1,58881
-11	-11	11	-11	1,47080
7	5	-1	5	1,38965
-9	-5	3	5	1,34383
3	1	9	-3	1,33233

Table 6 Elements of a vector *SL* and *COF* values for $k = 15$

sl_1	sl_2	sl_3	sl_4	*COF*
11	11	11	-11	1,56473
-11	-11	11	-11	1,53979
5	5	7	-1	1,35642
3	1	9	-3	1,34939
5	7	5	9	1,32918

easily noticed that differences between *COF* values are rather small and indicating the number of outliers on the basis of them may cause some difficulties.

The choice of *OF* factor do not change the structure of *COF* values for different instances, but it decides on the number of outliers detected by the algorithm. For example, if $k = 8$ and $OF = 9$ the system indicated only one outlier with $COF = 1,58881$ (see Table 5), while for $OF = 3$, the number of outliers indicated by the algorithm was equal to two as was expected, from the data structure. To obtain satisfying results, it is safer to choose a big number of neighbors.

The experiments were done, taking into account Euclidean distance function. Investigations presented in [54], where results for Euclidean and Hamming's functions were compared, showed that the choice of a distance function has no influence on the results of the *COF* algorithm. The same conclusion was drawn during research concerning cluster analysis of students according to learning styles dimensions, where Manhattan and Euclidean measures (see [55])were compared. As the result, in both stages of the presented method, the choice of Euclidean function as distance measure seems to be suitable.

Experiments concerning Farthest First Traversal clustering algorithm were done on the same sample clean data and mainly focus on the choice of the number of groups of students for construction of teaching paths. Investigations showed that, for the considered data set, number of K=4 clusters were the best. It was sufficient enough to divide students according to different characteristics and on the other hand not too big to control different teaching materials. Table 7 and Table 8 present cluster centres and number of instances for different K. Comparison of students groups centres and number of instances for K=3 and K=4, showed that in the second case the group of 5 students, active sensitive and global but without visual or verbal

Table 7 Cluster centres and number of instances for $K = 3$

sl_1	sl_2	sl_3	sl_4	Number of inst.
-1	1	-7	3	88
3	1	9	-3	12
-5	-11	-7	-7	23

Table 8 Cluster centres and number of instances for $K = 4$

sl_1	sl_2	sl_3	sl_4	Number of inst.
-1	1	-7	3	85
3	1	9	-3	11
-5	-11	-7	-7	22
-7	-11	1	5	5

preferences were found, these students were not distinguished for K=3 (see the last raw of the Table 8). Investigations showed that input parameter K should be chosen very carefully with tutor consulting.

The analysis of obtained clusters showed that some of students are characterized by more than one dominant learning style, what may reflect in different learner preferences. Using unsupervised classification for finding student groups with similar features allows for taking into account not only one learning style dimension, but combination of two or three of them, what may help during the process of differentiation of teaching materials. For example, in case of Computer Science students, most of them are strongly visual what can be easily noticed in column number 3 of Table 7 and Table 8 (more than 100 instances in each table), but that dimension cannot be the only one taken into account during students' profiles building.

4 Summary

In the chapter two case studies concerning application of data mining techniques for modeling users of educational portals is presented. The first one is focused on finding navigational patterns and clustering of users according to them, on the basis of historical data contained in log files. In the second one, there is examined application of cluster analysis on data collected from students, for dividing them according to learning styles' preferences. In both of considered cases, investigations showed that data mining algorithms indicated good performance for building groups of similar personal preferences.

The next step of research especially refering to data driven students' profiles, should concern connecting both of data sources and grouping users according to their dominant learning styles and navigational behaviors at the same time. Building a data cube containing both kinds of data may help in further application of data mining techniques for grouping users. In the next step, the system may be enhanced in information content and teaching paths' designers to automate the whole process.

References

1. van der Aalst, W.M.P., van Dongen, B.F., Herbst, J., Maruster, L., Schimm, G., Weijters, A.J.M.M.: Workflow mining: a survey of issues and approaches. Data Knowl. Eng. 47, 237–267 (2003)
2. Agrawal, R., Srikant, R.: Mining sequential patterns. In: Proc. of the 11th International Conference on Data Engineering (ICDE 1995), Taipei, Taiwan, pp. 3–14 (1995)
3. Beaudoin, M.F., Dwyer, M.: Learning or lurking? Tracking the "invisible" online student. Internet & Higher Educ. 5, 147–155 (2002)
4. Breuning, M.M., Kriegel, H.-P., Ng, R.T., Sander, J.: Identifying density-based local outliers. In: Proc. of ACM SIGMOD International Conference on Management of Data, pp. 93–104. ACM, Santa Barbara (2001)
5. Büchner, A.G., Mulvenna, M.D.: Discovering internet marketing intelligencethrough online analytical web usage mining. In: SIGMOD Record 27, pp. 54–61 (1998)
6. Cha, H.J., Kim, Y.S., Park, S.H., Yoon, T.B., Jung, Y.M., Lee, J.-H.: Learning styles diagnosis based on user interface behaviors for customization of learning interfaces in an intelligent tutoring system. In: Ikeda, M., Ashley, K.D., Chan, T.-W. (eds.) ITS 2006. LNCS, vol. 4053, pp. 513–524. Springer, Heidelberg (2006)
7. Chen, L., Bhowmick, S., Li, J.: COWES: Clustering web users based on historical web sessions. In: Lee, M.-L., Tan, K.-L., Wuwongse, V. (eds.) DASFAA 2006. LNCS, vol. 3882, pp. 541–556. Springer, Heidelberg (2006)
8. Cooley, R., Mobasher, B., Srivastava, J.: Web mining: information and pattern discovery on the world wide web. In: Proc. of the Ninth IEEE International Conference on Tools with Artificial Intelligence, pp. 558–567. IEEE, New York (1997)
9. Erinaki, M., Vazigriannis, M.: Web mining for web personalization. ACM T. Internet Technol. 3, 1–27 (2003)
10. Facca, F.M., Lanzi, P.L.: Mining interesting knowledge from weblogs: a survey. Data Knowl. Eng. 53, 225–241 (2005)
11. Felder, R., Brent, R.: Understanding student differences. J. Eng. Educ. 94, 57–72 (2005)
12. Felder, R.M., Silverman, L.K.: Learning and teaching styles in engineering education. Eng. Educ. 78, 674–681 (1988)
13. Fu, Y., Sandhu, K., Shih, M.Y.: Clustering of web users based on access patterns. In: Proc. of the 5th ACM SIGKDD International Conference on Knowledge Discovery & Data Mining. Springer, San Diego (1999)
14. Gilbert, J.E., Han, C.Y.: Adapting instruction in search of 'a significant difference'. J. Netw. Comput. Appl. 22, 149–160 (1999)
15. Graf, S., Kinshuk: Considering learning styles in learning managements systems: investigating the behavior of students in an online course. In: Proc. of the 1st IEEE Int. Workshop on Semantic Media Adaptation and Personalization, Athens (2006)
16. Han, J., Kamber, M.: Data Mining: Concepts and Techniques. Morgan Kaufman Publishers, San Francisco (2001)
17. He, z., Xu, X., Deng, S.: Discovering cluster-based local outliers. Pattern Recogn. Lett. 24, 1641–1650 (2003)

18. Hochbaum, D.S., Shmoys, D.: A best possible heuristic for the k-center problem. Math. Oper. Res. 10, 180–184 (1985)
19. Hodge, V.J., Austin, J.: A survey of outlier detection methodologies. Artif. Intell. Rev. 22, 85–126 (2004)
20. Honey, P., Mumford, A.: The Manual of Learning Styles. Peter Honey, Maidenhead (1986)
21. Huang, Y.M., Kuo, Y.H., Chen, J.N., Jeng, Y.L.: NP-miner: A real-time recommendation algorithm by using web usage mining. Knowl-Based Syst. 19, 272–286 (2006)
22. Jin, W., Tung, A.K., Han, J.: Mining top-n local outliers in large data bases. In: Proc. of the 7th ACM SIGKDD International Conference on Knowledge Discovery & Data Mining, pp. 293–298. ACM, San Francisco (2001)
23. Joshi, K.P., Joshi, A., Yesha, Y.: On using a warehouse to analyze web logs. Distrib. Parallel Dat. 13, 161–180 (2003)
24. Karger, D.R., Qan, D.: Prerequisites for a personalizable user interface. In: Proc. of Intelligent User Interface 2004 Conf. Ukita (2004)
25. Knorr, E.M., Ng, R.T., Tucakov, V.: Distance-based outliers: algorithms and applications. VLDB J. 8, 237–253 (2000)
26. Kolb, D.A.: Experiental Learning: Experience as a Source of Learning and Development. Prentice-Hall, Englewood Cliffs (1984)
27. Lee, M.: Profiling students' adaptation styles in web-based learning. Comput. Educ. 36, 121–132 (2003)
28. Liu, R.-L., Lin, W.-J.: Mining for interactive identification of users' information needs. Inform. Syst. 28, 815–833 (2003)
29. Lu, J., Yu, C.S., Liu, C.: Learning style, learning patterns, and learning performance in a WebCT-based MIS course. Inform. Manage. 40, 497–507 (2003)
30. Luotonen, A.: The common logfile format (1995), `http://www.w3.org/pub/WWW/Daemon/User/Config/Logging.html`
31. Mobasher, B., Cooley, R., Srivastava, J.: Automatic personalization based on web usage mining. Commun. ACM 43, 142–151 (2000)
32. Morzy, T., Wojciechowski, M., Zakrzewicz, M.: Web users clustering. In: Proc. of the 15th International Symposium on Computer and Information Sciences, Istanbul, Turkey, pp. 374–382 (2000)
33. Pabarskaite, Z., Raudys, A.: A procss of knowledge discovery from web log data: systematization and critical review. J. Intell. Inf. Syst. 28, 79–104 (2007)
34. Paliouras, G., Papatheodorou, C., Karkaletsis, V., Spyropoulos, C.D.: Clustering the users of large web sites into communities. In: Proc. of International Conference on Machine Learning ICML, Stanford, California, pp. 719–726 (2000)
35. Pei, J., Han, J., Mortazavi-asl, B., Zhu, H.: Mining access patterns efficiently from web logs. In: Terano, T., Liu, H., Chen, A.L.P. (eds.) PAKDD 2000. LNCS, vol. 1805, pp. 396–407. Springer, Heidelberg (2000)
36. Pierrakos, D., Pakouras, G., Papatheodorou, C., Spyropoulos, C.D.: Web usage mining as a tool for personalization: a survey. User Model. User-Adap. 13, 311–372 (2003)
37. Ramaswamy, S., Rastogi, R., Shim, K.: Efficient algorithms for mining outliers from large data sets. In: Proc. of ACM SIGMOD International Conference on Management of Data, pp. 427–438. ACM, Dallas (2000)
38. Romero, C., Ventura, S. (eds.): Data Mining in E-learning. WIT Press, Southampton (2006)
39. Rovai, A.P.: The relationships of communicator style, personality-based learning style and classroom community among online graduate students. Internet & Higher Education 6, 347–363 (2003)

40. Saarikoski, L., Salojärvi, S., del Corso, D., Ovein, E.: The 3DE: an environment for the development of learner-oriented customized educational packages. In: Proc. of ITHET 2001, Kumamoto, Japan (2001)
41. Soloman, B.A., Felder, R.M.: Index of Learning Style Questionnaire (2007), `http://www.engr.ncsu.edu/learningstyles/ilsweb.html` (Cited November 30, 2007)
42. Srikant, R., Yang, Y.: Mining web logs to improve website organization. In: Proc. of WWW 2001, pp. 430–437. ACM, Hong Kong (2001)
43. Stash, N., Cristea, A., de Bra, P.: Authoring of learning styles in adaptive hypermedia: Problems and solutions. In: Proc. of WWW 2004, pp. 114–123. ACM, New York (2004)
44. Su, C.M., Tseng, S., Jiang, M.F., Chen, J.C.: A fast clustering process for outliers and reminder clusters. In: Zhong, N., Zhou, L. (eds.) PAKDD 1999. LNCS (LNAI), vol. 1574, pp. 360–364. Springer, Heidelberg (1999)
45. Tang, J., Chen, Z., Fu, A.W., Cheung, D.W.: Enhancing effectiveness of outlier detections for low density patterns. In: Cheng, M.-S., Yu, P.S., Liu, B. (eds.) PAKDD 2002. LNCS, vol. 2336, pp. 535–548. Springer, Heidelberg (2002)
46. Teo, H.H., Chan, H.C., Wei, K.K., Zhang, Z.: Evaluating information and community adaptivity features for sustaining virtual learning communities. Int. J. Hum.-Comput. St. 59, 671–697 (2003)
47. Triantafillou, E., Pomportsis, A., Georgiadou, E.: AES-CS: Adaptive Educational System base on cognitive styles. In: Proc. AH Workshop, Malaga, pp. 10–20 (2002)
48. Viola, S.R., Graf, S., Kinshuk, Leo, T.: Investigating relationships within the Index of Learning Styles: a data driven approach. Interactive Technology & Smart Education 4, 7–18 (2007)
49. Wang, W., Zaïane, O.R.: Clustering web sesions by sequence alignment. In: Proc. of 13th International Conference on Database and Expert Systems Applications DEXA, Aixen Provence, France, pp. 394–398 (2002)
50. Xiao, J., Zhang, Y.: Clustering of web users using session-based similarity measures. In: Proc. of the 2001 International Conference on Computer Networks and Mobile Computing ICCNMC 2001, pp. 223–228 (2001)
51. Zaïane, O.R., Xin, M., Han, J.: Discovering web access patterns and trends by applying OLAP and data mining technology on web logs. In: Proc. of Advances in Digital Libraries Conference (ADL 1998), Santa Barbara, CA, pp. 19–29 (1998)
52. Zakrzewska, D.: Finding navigational patterns of web users by log file mining (in Polish). In: Kozielski, S., Małysiak, B., Kasprowski, P., Mrozek, D. (eds.) Databases New Technologies. Architecture, Formal Methods and Advanced Data Analysis, Wydawnictwo Łacznosci i Telekomunikacji, Warszawa, pp. 393–400 (2007)
53. Zakrzewska, D.: On using OLAP mining for finding web navigation patterns. In: Borzemski, L., Grzech, A., Światek, J., Wilimowska, Z. (eds.) Information Systems Architecture and Technology. Information Technology and Web Engineering: Models, Concepts & Challenges, Oficyna Wydawnicza Politechniki Wrocławskiej, Wrocław, pp. 109–116 (2007)
54. Zakrzewska, D., Osada, M.: Outlier analysis for financial fraud detection. In: Dramiński, M., Grzegorzewski, P., Trojanowski, K., Zadrożny, S. (eds.) Issues in Intelligent Systems. Models and Techniques, pp. 291–306. AOW EXIT, Warszawa (2005)
55. Zakrzewska, D., Ruiz-Esteban, C.: Cluster analysis for students profiling. In: Byczkowska-Lipińska, L., Szczepaniak, P.S., Niedźwiedzińska, H. (eds.) Proc. of the 11th International Conference on System Modelling Control, pp. 333–338. AOW EXIT, Warszawa (2005)
56. `http://www.cs.washington.edu/ai/adaptive-data`

Frequency Domain Methods for Content-Based Image Retrieval in Multimedia Databases

Bartłomiej Stasiak and Mykhaylo Yatsymirskyy

Abstract. Content-based image retrieval is an important application area for image processing methods associated with computer vision, pattern recognition, machine learning and other fields of artificial intelligence. Image content analysis enables us to use more natural, human-level concepts for querying large collections of images typically found in multimedia databases. Out of the numerous features proposed for image content description those based on frequency representation are of special interest as they often offer high levels of invariance to distortions and noise. In this chapter several frequency domain methods designed to describe different aspects of an image, i.e. contour, texture and shape are discussed. Current standards and database solutions supporting content-based image retrieval, including SQL Multimedia and Application Packages, Oracle 9i/10g interMedia and MPEG-7, are also presented.

1 Introduction

The evolution of consumer electronics products over the past decades is inevitably associated with changes of the processed data characteristics. Numerical and textual operations constitute no longer the core of typical applications of a personal computer, yielding to multimedia content processing. This tendency is reflected in the need of development of new database solutions in which images, video sequences or audio recordings would be efficiently stored and retrieved in a similar manner as in the case of traditional alphanumeric data.

Bartłomiej Stasiak
Institute of Computer Science, Technical University of Lodz, ul. Wolczanska 215, 90-924 Lodz, Poland
e-mail: `basta@ics.p.lodz.pl`

Mykhaylo Yatsymirskyy Institute of Computer Science, Technical University of Lodz, ul. Wolczanska 215, 90-924 Lodz, Poland
e-mail: `jacym@ics.p.lodz.pl`

D. Zakrzewska et al. (Eds.): Meth. and Support. Tech. for Data Analys., SCI 225, pp. 137–166.
springerlink.com

The complexity of tasks required from modern image, audio and video processing tools results not only from the sheer data dimensionality growth, but also from the specific structure and form of information contained in these media. Extending the existing methodology of storing and retrieving data by database management systems (DBMS) to the form of multimedia DBMS (MMDBMS) we encounter several problems related to either the quantity or quality of the processed data. Some of the issues, such as concurrency, transaction management, versioning, backup and recovery procedures may be effectively solved on the basis of the existing systems, e. g. using the binary large object type (BLOB) for storing multimedia content. The rapidly increased demand for disk space and transmission bandwidth may enforce changes in hardware configuration, including external storage media or streaming technologies application. The most significant challenge, however, is the need of construction of information retrieval infrastructure enabling the users to specify their queries based directly on the multimedia content [1].

The search criteria used in multimedia databases fall into two general categories: based on external metadata and content-based. In the first case we rely on some additional information associated with each object in the searched collection, such as filename, data format, date, size or manually added textual description. For this kind of search traditional techniques based on alphanumeric data processing may be easily applied. The content-based approach is a more difficult, yet very promising and potentially robust alternative in which the query describes the actual content of the multimedia object [2].

The most challenging problem in content-based image retrieval (CBIR) is the semantic gap between low-level description possibilities achievable by image analysis methods and the high degree of abstraction expected by the users [3], [4]. The human-level concepts used to construct a query are often far beyond what can be directly expressed in terms of color distribution, shape and texture characteristics. Moreover, many crucial operations such as image segmentation or edge detection may be vulnerable to lighting conditions, partial occlusions, noise or changes in the spatial localization of image components. For these reasons frequency domain methods are an interesting alternative to direct image analysis in the time domain as they usually offer more general and noise-immune approach.

In this chapter we will cover several frequency domain methods for content-based image retrieval from multimedia databases. We will also present and discuss other known solutions, prototype systems and current standards supporting image content analysis.

2 The Basics of Automatic Image Retrieval and Classification

The general approach to multidimensional data searching, recognition and classification is based on feature extraction. Images, being examples of highly dimensional entities, need to be represented by small portions of data, containing only the most important, condensed information enabling to deduce the image content. Such a representation is referred to as an image *signature* which should be comprehensive (i.e. it should completely and properly represent the image), compact (i.e. having low storage requirements) and computationally effective. The process of constructing the signature based on some carefully selected visual features is actually the most crucial stage of any CBIR system, as ill-formulated features definition makes all other efforts to properly recognize or classify the image futile [5], [6]. The general types of signatures applied to describe image content include single feature vector, region-based signature and a summary of local feature vectors [4].

The first approach is typical for general classification problems in which we build a direct mapping between the real objects domain and the feature space. In the case of image classification we search for a global representation describing the content of the whole image, such as e.g. global shape descriptors or color histogram computed for all the image pixels.

The single feature vector construction was the most common approach applied in the last decades of the twentieth century. The dominating trend in the more recent research is to build local descriptions of image regions extracted during the preceding segmentation process. It is now understood that this is a more promising way to reduce the semantic gap on the basis of a higher-level analysis of image structure and the relations between its components. In this approach the image signature is build as a weighted sum of the vectors representing the regions. The challenge involved here is how to design a reliable segmentation procedure, which still remains an open problem in the general case. Several methods have been developed to reduce the sensitivity of the region-based image retrieval techniques to the segmentation results (e.g. [7], [8], [9]).

The approach based on local feature vectors requires that characteristics of small adjacent image blocks, possibly corresponding to the neighborhood of every image pixel, are computed and summarized. The summary used for the signature construction may be obtained in several ways on the basis of the distribution of the analyzed feature over the whole image area.

Once the signature has been constructed, it may be used to assess similarity between images on the basis of some kind of similarity measure. Several solutions to this problem exist including matching methods based on visual features, structural features, salient features or semantic-level matching [3], to name just a few. The matching techniques are also dependent on database structure, user interface organization and system requirements. Two typical search possibilities are: finding images similar to the user-specified example and determining the image class it belongs to. In the second case some classification methods (e.g. k nearest neighbors, neural networks [10] or support vector machines [11]) must be additionally applied.

3 Solutions and Standards Supporting Content-Based Image Retrieval

The growing demand for multimedia-aware database systems triggered the development of various prototype systems and made the need for standardization in this field clear. One of the main problems was that the capabilities of the previously sufficient query languages were obviously too limited for multimedia content-based retrieval. The solutions followed either of two ways - new multimedia-oriented query languages and database systems were proposed or the existing technology was extended to rise to the new challenges.

New solutions natively implementing operations on multimedia content are generally easier to adjust to the specific needs imposed by the nature of the stored data. The efficiency of such systems is also usually better. The disadvantages of this approach include the need of implementing from scratch all the basic database functionality, such as concurrency, transactions, data integrity control and relational mechanisms. Some of the solutions designed to support content based image retrieval from databases are briefly described in the following part of this section. Current standards and their implementations are presented in subsections 3.1, 3.2 and 3.3.

Refined entity-relationship model for representing image content forms the basis of SCORE [12] (a System for COntent based REtrieval of pictures). The entities correspond to the real-world objects depicted in an image while the relations represent either interactions or spatial dependencies between them. The iconic user interface enables the user to specify queries in a natural, intuitive way and the results are presented both in the form of a picture and its symbolic representation. The system supports a method (selective relevance feedback) for defining new, more detailed queries on the basis of the previous results which may be partially or entirely reused. The system can also model the generalization-specialization relationships between objects and some properties of spatial dependencies, such as transitivity of directional spatial relationships.

SMDS (Structured Multimedia Database System) proposed in [13] is based on a theoretical approach to multimedia content and structure description. The authors introduce a mathematical model for *media-instance* which is a concept representing various kinds of physical media (such as video, audio, documents, etc.). The model consists of "states" (corresponding to e.g. video frames or audio tracks), "features" and relationships between them. The media-instance abstraction is used to construct a general purpose logical query language, based on SQL syntax and offering additional functionality, such as functions `FindType`, `ObjectWithFeature` or `FindFeaturesinObject`, for multimedia data querying.

MOQL (Multimedia Object Query Language), introduced in [14], comprises a set of Object Query Language (OQL) extensions intended to support multimedia handling functionality. One of the main objectives of the MOQL construction was its generality and independence of any particular media type or application. The extensions are mostly connected to the WHERE clause and they include three predicate expressions: *spatial_expression*, *temporal_expression* and *contains* predicate. The *spatial_expression* may include objects (e.g. points, lines, circles), functions (e.g. length, area, intersection) and predicates (e.g. cover, disjoint, left). The *temporal_expression* deals with temporal relationships e.g. in a video sequence. The *contains* predicate is used to determine whether an object of interest (salient object) is present in a given media instance.

CSQL (Cognition and Semantics-based Query Language) was introduced in [15] as a part of SEMCOG image database system (SEMantics and COGnition-based image retrieval). This SQL-based language enables to specify queries on the basis of object-level information including both visual and semantic, cognition-related object characteristics. Three types of entities are used in CSQL: semantic entities, image entities and dual entities. All three of them are further classified into compound and atomic. Semantic entity has only semantic property and may consist of a semantic term or a semantic description of an image or an object. Image entity is an entity with only image property referring to an image or image object. Dual entities are polymorphic entities with both image property and semantic property. Several new predicates, e.g. `is`, `contains`, `to_the_right_of`, were also introduced to support selection criteria construction.

3.1 SQL Multimedia and Application Packages (SQL/MM)

The SQL standard was initially planned to incorporate the new multimedia-related functionality but it quickly appeared that the problems of data incompatibility and conflicting keywords would make the smooth integration impossible. One of the simple examples of conflicting semantics was the keyword CONTAINS used both in text operations and for determining the spatial relationship between objects in an image.

A new standard based on libraries using object types defined in SQL99 specification was therefore developed. SQL Multimedia and Application Packages, known as SQL/MM is an ISO/IEC standard supporting multimedia content storage and management [16], [17]. It comprises five parts:

- Part I: *Framework* specifies how the SQL object types are used and what are the relations between the remaining parts of the standard.
- Part II: *Full-Text* defines data types for storing big text objects and specifies the methods for efficient searching and other text operations.
- Part III: *Spatial* defines data types supporting spatial information processing. The support of zero-dimensional objects (points), one-dimensional (lines) and two-dimensional (planar figures) is included.
- Part IV: *General Purpose Facilities* was intended to comprise some general-purpose mathematical procedures, but it is currently not developed.

- Part V: *Still image* specifies data types supporting image processing and management.
- Part VI: *Data mining* defines data types specialized for data exploration tasks.

Part V of the standard (*Still image*) supports mechanisms for storing, retrieving and manipulating images in a database. The SQL99 type, called `SI_StillImage` is used for this purpose. The image is stored in the form of a two-dimensional collection of pixels as a BLOB type along with additional information about the dimensions, color space and format. The available formats depend on the implementation, typically JPEG, GIF and TIFF are supported. The image manipulation possibilities include scaling, cropping and rotation. A thumbnail representation of `SI_StillImage` instances may be also easily generated.

Several object types exist for representation of various low-level features of an image. The available types include three color descriptors (`SI_AverageColor`, `SI_ColorHistogram` and `SI_PositionalColor`) and texture descriptor (`SI_Texture`). A separate type named `SI_FeatureList` enables to group single image descriptors and to specify the weight coefficient associated with each of them to improve the precision and effectiveness of searching and retrieval.

Each image content property may be retrieved by a special function, such as `SI_findAvgColor` or `SI_findTexture` (returning `SI_AverageColor` or `SI_Texture` instance respectively). The descriptor objects in turn have a `SI_Score` method defined for calculating a distance from other images in the sense of a given feature. The following example demonstrates how to select images from table Products which have the average color similar to a queryImage:

```
SELECT * from Products
where SI_findAvgColor(queryImage).
SI_Score(product_photo) < 1.5;
```

3.2 Oracle 9i/10g interMedia

The functionality described in the SQL/MM, concerning content-based image retrieval and manipulation, is present in many database management systems (e.g. DB2). However, they are usually not fully SQL/MM-compliant on the interface level. The Oracle interMedia [18] included in Oracle 9i is an example of a multimedia database subsystem which enables to store images, extract signatures and specify CBIR queries but the object types and methods it defines are different from the SQL/MM *Still image* specification. The Oracle 9i interMedia functionality is accessible mainly through four object types defined in ORDSYS scheme: ORDAudio (audio objects storing and management), ORDVideo (video objects storing and management), ORDImage (still images storing and management) and ORDDoc (representing any media data). In the following example we will show how to apply ORDImage object to construct a simple query based on image visual appearance criteria.

In order to perform content-based queries on a collection of images, which are stored in a table column of type `ORDSYS.ORDImage`, an additional column of type `ORDSYS.ORDImageSignature` must be created. The images can be inserted into the table as follows:

```
INSERT INTO wallpapers
  (wallpaper_id, wallpaper_img, wallpaper_signature) VALUES
  (1, ORDSYS.ORDImage.init('FILE', 'IMG_DIR','straw.jpg'),
  ORDSYS.ORDImageSignature.init());
```

and the signatures must be explicitly generated with `generateSignature` method:

```
DECLARE
  wp_img ORDSYS.ORDImage;
  wp_sig ORDSYS.ORDImageSignature;
BEGIN
  SELECT wp.wallpaper_img, wp.wallpaper_signature
    INTO wp_img, wp_sig FROM wallpapers wp
    WHERE wp.wallpaper_id = 1 FOR UPDATE;
  wp_sig.generateSignature(wp_img);
  UPDATE wallpapers wp SET wp.wallpaper_signature = wp_sig
    WHERE wallpaper_id = 1;
END;
/
```

The `ORDSYS.IMGSimilar` operator may be applied to find matching images in the database, as in the following query in which we search for all images similar to the first one:

```
SELECT wp.wallpaper_id FROM wallpapers wp WHERE
  ORDSYS.IMGSimilar(wp.wallpaper_signature,
  (SELECT q.wallpaper_signature FROM wallpapers q
    WHERE q.wallpaper_id = 1),
  'color="0.3" texture="0.5" shape="0.1"
  location="0.1"', 20) = 1;
```

On the call to `ORDSYS.IMGSimilar` the signatures to compare must be specified as well as the value of threshold and weights assigned to different visual attributes. Four attributes are allowed, i.e. color, texture, shape and location and their weights may vary from 0.0 (ignored attribute) to 1.0 (most relevant).

The weighted sum of the attributes is compared against the threshold value chosen from the range 0..100, yielding the return value of '1' for similar images, '0' otherwise.

The compliance with SQL/MM *Still image* specification is provided by the newer edition Oracle 10g which specifies also the set of types and methods required by the SQL/MM standard in addition to the ORDSYS objects known from the previous version. It should be noted that some functionality offered by interMedia, including shape and localization descriptors accessible via ORDImageSignature type, is an extension not covered by the SQL/MM specification.

Multimedia objects managed by interMedia may be:

- stored internally in the database in the form of binary objects (BLOB),
- stored in the local filesystem and accessed with BFILE type,
- accessed from external localization specified by a URL,
- accessed from external streaming servers (this option may need installing additional components).

Using BLOB type enables to apply access control and transactional mechanisms leading to increased integrity protection of the image data. On the other hand the external localization makes the import of existing, large multimedia collections into the Oracle environment much easier. The details of the actual localization of data are hidden in a special ORDSource type, which is a component of all multimedia data types (ORDAudio, ORDVideo, ORDImage and ORDDoc). This additional abstraction layer unifies the access to binary data, irrespective of the actual type of media they represent.

3.3 MPEG-7

The Moving Picture Experts Group is primarily known for the popular standards of audio/video data coding (MPEG-1, MPEG-2, MPEG-4). In contrast, MPEG-7 does not describe a new codec or data format. This ISO/IEC standard "describes how to describe" the data, i.a. it defines methods and tools for multimedia objects (MM) analysis designed to facilitate annotating and content-based searching and retrieval from multimedia resources [19].

The tools for visual content analysis, described in the third part of the MPEG-7 standard [20], cover the following basic areas: Color, Texture, Shape, Motion, Localization, and Face Recognition. Additionally, there are also five basic structures defined: Grid Layout, 2D-3D Multiple View, Spatial 2D Coordinates, Time Series and Temporal Interpolation, where the last two are used for video sequences description. In the following section we will briefly summarize the visual descriptors most suitable for still image content analysis.

There are three basic structures relevant to still images description:

- *Grid Layout* enables to describe each element of a set of equally sized rectangular image regions independently.

- *2D-3D Multiple View Descriptor* enables to represent a 3-dimensional object by its 2-dimensional descriptions corresponding to possibly different view angles.
- *Spatial 2D Coordinates Descriptor* offers a way to make other image content descriptors independent on the image scale changes.

There are seven color descriptors: Color Space, Color Quantization, Dominant Colors, Scalable Color, Color Layout, Color-Structure and Group of Frames/Group of Pictures color descriptor (GoF/GoP):

- *Color Space* allows to choose the color model used (e.g. RGB, HSV, YCrCb, monochrome).
- *Color Quantization* defines the number of color bins obtained after quantization.
- *Dominant Color* is used in combination with color quantization to represent image regions by means of the dominant colors description.
- *Scalable Color* defines the image histogram (in HSV color space), which is scalable in terms of bin numbers and the representation accuracy. The histogram is encoded by Haar transform.
- *Color Layout Descriptor* describes the spatial distribution of colors. The image is divided into blocks, each represented by its dominant color, to form a thumbnail representation which is then transformed by the discrete cosine transform. A few low frequency coefficients are then used to build the descriptor. Its compactness enables to reach low computational cost of the matching process. It is worth noting that this is the only descriptor providing the mechanism in which the user may submit hand-written sketch queries.
- *Color-Structure* is based on the approach similar to histogram analysis with the difference lying in the possibility to distinguish images with the same histograms by taking into account the local structure of pixel neighborhoods.
- *Group of Frames/Group of Pictures Color* is an extension of the Scalable Color Descriptor to a group of frames or still images.

The Illumination Invariant Color Descriptor, based on four color descriptors defined in [20]: Dominant Color, Scalable Color, Color Layout, and Color Structure, was added in the newer version of the standard (MPEG-7 Visual Amendment 1) to increase the independence of lighting conditions.

The MPEG-7 standard specifies three texture descriptors: Edge Histogram, Homogeneous Texture and Texture Browsing:

- *Edge Histogram* distinguishes five types of edges (four directional edges and one non-directional) which are used for local histograms construction. The local histograms can be then aggregated into global or semi-global ones.
- *Homogeneous Texture Descriptor* is designed to describe image regions filled with texture of approximately constant characteristics. The descriptor is based on a bank of orientation- and scale-tuned Gabor filters.

- *Texture Browsing Descriptor* is constructed in the similar way to the Homogeneous Texture Descriptor. It is intended for browsing type applications as it is represented on much smaller number of bits.

There are three shape Descriptors defined in MPEG-7: Region Shape, Contour Shape, and Shape 3D.

- *Region Shape* describes the shape of an object defined by set of one or more, possibly disjoint regions. The regions may contain holes which do not belong to the object.
- *Contour Shape Descriptor* is based on the curvature scale space model (CSS). In this case the interior of the object is neglected and only its boundary is taken into account.
- *Shape 3D Descriptor* provides shape description of 3-dimensional mesh models, by the analysis of some local attributes of the 3D surfaces.

The MPEG-7 standard includes also a specialized *Face Recognition Descriptor* intended exclusively for face images indexing. The descriptor is constructed from images of size 46x56 on the basis of several methods including Fourier analysis and linear discriminant analysis (LDA).

4 Time Domain Methods – Overview

In this section the existing time domain methods for image feature extraction are described and briefly summarized. The term "time domain", conventionally used by analogy to one-dimentional time series analysis, is equivalent to "spatial domain" in this context.

The discrimination of the image analysis methods into time- and frequency-based is dependent on the type of the features extracted. Some features are in a more or less natural way better described in either of the two domains and for some both approaches may yield a satisfying result. There are also cases in which the feature based on spatial characteristics is additionally transformed into frequency representation (e.g. MPEG-7 Scalable Color or Color Layout Descriptor).

Color

Color is the feature most eagerly applied to characterize images from the very beginning of the CBIR history. Although there are many methods of image signature construction based on monochrome representations, which may be the right approach in some cases (e.g. the color of man-made objects is sometimes meaningless), it is nevertheless obvious that ignoring colors reduces the available information substantially. The most prominent ways of using the color information include color moments and color histograms.

Color moments have been successfully applied in many systems including QBIC [21], as they offer a very compact image representation. In this method image color distribution is treated as probability distribution and described by standard statistical measures. The first three central moments (mean, standard deviation and skewness) are usually considered for the three channels of HSV, CIE (L*, a*, b*) or CIE (L*, u*, v*) color space [22], [23].

Color histogram is an effective form of color representation which is easy to compute and flexible in terms of the possibility of choosing an arbitrary number of histogram bins. It has been widely applied for image indexing and retrieval [24], [25], [26] due to its advantages including the relatively high level of invariance to affine transforms and partial occlusions. On the other hand, histogram contains the information about the pixel quantities, ignoring their localization. It is therefore not too difficult to find two totally different pictures with similar histograms, especially in large image databases. Computing local histograms for image regions obtained during segmentation or partitioning is a possible solution to this problem.

Other methods of incorporating spatial information into histogram-based descriptors include color coherence vector (CCV) [27] and color correlogram [28]. The problem of invariant color representation under varying conditions involving illumination changes, specular reflection, as well as geometrical and reflective properties of surfaces have been studied more recently in [29], [30], [31].

Texture

Texture analysis is the research area in which a plethora of methods, mostly based on some kind of frequency representation, have been proposed [32], [33], [34], [35], [36], [37], [38]. This variety is in a way connected to the existence of many specific categories of images (e.g. aerial or medical imaging) for which the texture classification and its semantic interpretation are of great practical significance. It must be noted that, due to the specific characteristics of textures and methods applied for their analysis, a clear distinction between time- and frequency-domain methods is not always possible. One of the existing divisions distinguishes between structural and statistical methods [39].

The methods of texture features extraction based on structural approach are basically defined in the time-domain. These methods try to describe texture components in the local spatial context using such tools as morphological operations and adjacency graphs. This class of techniques is generally best applicable to images with regular textures.

Statistical methods are based on statistical moments of gray-level texture histograms or on the response of oriented band-pass filters using Fourier, wavelet or Gabor transforms [40], [41], [42], [43]. Selected output coefficients of those transforms or their statistics are taken to form texture features. Other techniques proposed for texture analysis include Tamura features [44], Wold decomposition [45], multi-resolution simultaneous auto-regressive model [46], PCA-based methods and a class of algorithms based on co-occurrence matrices.

Shape

Shape may be seen as a higher-level concept with respect to image segmentation. Color and texture help to find separate regions, while shape analysis is usually dependent on the segmentation results. Global shape representations were also used in practice, e.g. in [21], [47], [48], nevertheless local shape descriptors seem to dominate in more recent applications [49], [50], [51], [52]. One of the main objectives in the problems of shape definition and construction of matching algorithms is how to achieve invariance to affine transforms and other distortions which may dramatically change the visual appearance of the shape without altering their semantics.

The description of a shape may be based either on the whole region or on its boundary. The first approach include the classical techniques based on statistical moments, used to define moment invariants to rotation, translation and scaling [53], [56]. The boundary-based methods generally need an additional step of contour tracking (e.g. with active contours approach [54], [55]). The boundary may be then represented in the form of a sequence of points on a plane.

The boundary-based methods may be further divided into structural and non-structural. The structural approach to boundary-based shape description breaks the contour into segments or *primitives* which are used to build a kind of string or tree representation [49], [57]. Non-structural approaches treat the boundary as a whole and define its signature on the basis of some characteristic points (e.g. curvature zero-crossing points in CSSD [58]) or frequency-domain coefficients.

5 Frequency Domain Methods

In the field of signal processing the frequency and time domain representations are equivalent and they complement each other. The usefulness of either of them is dependent on the type of the signal and the purpose of the analysis. For example, processing audio content, which is a typical one-dimensional signal, most often relies heavily on frequency domain techniques.

Images, being the typical examples of two-dimensional signals, are quite different. The way in which human brain perceives and analyzes images seems to have little in common with their spectral content. The features we are most sensitive to are colors, shapes or spatial relationships between objects and no direct, natural method of their description in the frequency domain exists. The main reason for the prevalence of time, or spatial domain methods in the practice of content-based image recognition and retrieval is therefore the more straightforward and similar to the human vision approach they offer.

Nevertheless, as our knowledge about the functioning of human senses and the central nervous system is still incomplete, the apparent analogy of time-domain techniques proposed for image signature extraction and the simplistic idea of image analysis methods presumably applied by the human visual system may be illusory. Apart from complete higher level semantic interpretation of the scene, which still remains beyond the reach of CBIR systems, even the low-level factors, such as noise, varying lighting conditions or different angle of view may thwart the effort to reliably describe the image content. Several classes of features described in the previous section are likely to be susceptible to these types of distortions. Noise may influence texture analysis methods and local feature descriptors. A change of light sources location may totally unsettle the image color layout and no single global color-based feature would reliably describe its content. A change of the viewing point reflected in affine transformations and other image modifications, such as partial occlusions of the depicted objects, may significantly alter these image characteristics which are crucial for region based techniques, shape analysis and semantic description methods.

5.1 Inspiration

The problems of low-level image description may be seen as the consequence of the existence of the sensory gap defined in [3] as the discrepancy between the object in the world and the information in a (computational) description derived from a recording of the scene. The question arises whether it is possible to build a uniform representation of an image that would be immune to a certain degree of both local and global distortions without having to inspect their every possible source.

We are interested in a *global* description, not necessarily in the sense of analyzing the image as a whole, but rather in the sense of avoiding extraction of many very specialized features. The analysis of image parts or regions may be also performed but it should be highly independent on the quality of the preceding segmentation. In real-world applications the boundary of a segmented region containing an object may only roughly correspond to the actual shape of the object, which will lead to recognition failure for most of methods dependent on e.g. contour-tracking algorithms. A more robust approach, which will be covered in detail in this section, is based on the whole image or region analysis performed in the frequency domain.

Spectral analysis of an image, while apparently lacking some useful features, e.g. the ability to easily describe the spatial location of object components, may at the same time be attractive in the sense of sensory gap reduction possibilities. Both local and global image characteristics are easy to select and filter if necessary. Certain kinds of discrepancies commonly found between the similar pictures in real-world image collections are easy to recognize and circumvent. Additionally, transforming an image to the frequency domain enables to substantially reduce the redundancy of the data due to the decorrelation properties of the typically applied Fourier or cosine transforms. It should be noted that the reduction ratio may be easily adjusted, thus enabling to control the compactness and precision of the obtained representation. For some approaches it might be even possible to reconstruct the image from the signature with arbitrary precision [59], which is clearly impossible in case of histograms, statistical moments or most of other time-domain techniques.

Due to these advantages, the spectral methods are also applied to other, more specific areas of image content analysis summarized in section 4, i.a. for constructing effective texture or contour descriptors. Existing methods of Fourier transform application for boundary-based shape description and texture analysis will be presented in sections 5.2 and 5.3 respectively. The main contribution of this chapter i.e. the method of obtaining general image descriptor, based on Fourier-Mellin transform, and its applications to content-based image recognition will be presented in section 5.4.

5.2 Boundary-Based Shape Descriptors

A closed contour of an object may be represented as a periodic complex function $z(s) = x(s) + iy(s)$, $s \in \Re$ with period T where $(x(s), y(s))$ correspond to the spatial coordinates of a contour point. The contour is characterized by its turning function [39] defined as:

$$\theta(s) = \arctan\left(\frac{y'(s)}{x'(s)}\right) ,$$

$$y'(s) = \frac{dy(s)}{ds} , \tag{1}$$

$$x'(s) = \frac{dx(s)}{ds} .$$

This formulation is, however, not invariant to rotation nor to the choice of the starting point, which makes it difficult to apply for shape similarity comparison. To overcome this problem the frequency representation, based on Fourier descriptors, may be considered instead.

The first step of the computation involves the choice of the contour representation ensuring shift invariance. There exist several possibilities here, i.a. complex coordinate function, curvature representation and centroid distance.

The complex coordinate function is a shift-invariant form of the function $z(s)$:

$$Z(s) = (x(s) - x_c) + i\,(y(s) - y_c) , \tag{2}$$

where $(x_c,\ y_c)$ are the coordinates of the centroid:

$$(x_c,\ y_c) = \left(\frac{1}{T}\int_0^T x(s)ds,\ \frac{1}{T}\int_0^T y(s)ds\right) .$$

The curvature $C(s)$ is defined as a derivative of the turning function (1):

$$C(s) = \frac{d\theta(s)}{ds} , \tag{3}$$

The centroid distance is simply the euclidean distance between a point on the contour and the centroid of the analyzed object:

$$r(s) = \sqrt{(x(s) - x_c)^2 + (y(s) - y_c)^2} , \tag{4}$$

where (x_c, y_c) are the coordinates of the centroid.

Choosing one of the forms (2 - 4) influences the next computational steps. In particular, in formula (2) a sequence of complex numbers is obtained which means that no Fourier spectrum symmetry, occurring in the case of (3) or (4) application, may be expected. In the following part the latter representation will be used (for a comparison between different methods of contour analysis see [60]).

A more appropriate, discrete form of (4) is defined for a sequence of N points sampled along the contour:

$$r(n) = \sqrt{(x(n) - x_c)^2 + (y(n) - y_c)^2} , \tag{5}$$

where $n = 0, 1, ..., N - 1$ and (x_c, y_c) are the coordinates of the centroid:

$$x_c = \frac{1}{N}\sum_{n=0}^{N-1} x(n), \; y_c = \frac{1}{N}\sum_{n=0}^{N-1} y(n) . \tag{6}$$

The Fourier descriptors are computed on the basis of (5) with the one - dimensional discrete Fourier transform (DFT):

$$R_N(k) = DFT_N\{r(n)\} = \frac{1}{N}\sum_{n=0}^{N-1} r(n)e^{-i\frac{2\pi}{N}kn} , \tag{7}$$

for $k = 0, 1, ..., N - 1$.

In practice, the exact number of the coefficients $R_N(k)$ is much smaller than N. The coefficients corresponding to higher frequencies are generally related to very fine details or noise and, as such, they are considered redundant in most classification tasks. Only the first M output values of (7) containing the most important information about the shape are therefore used.

The issue of invariance to scaling, rotation and the choice of the starting point may be easily addressed in the frequency domain. In the commonly applied approach the first Fourier output coefficients, i.e. the modulus of $R_N(0)$ and the argument of $R_N(1)$, are used to normalize the whole representation, yielding the invariance to scaling and rotation/starting point choice respectively. The components of the normalized vector of Fourier descriptors $\{\overline{R}(k) : k = 1, 2, ..., M-1\}$ are defined as:

$$\overline{R}(k) = \frac{|R_N(k)|}{|R_N(0)|} \cdot e^{-i(\arg(R_N(k)) - \arg(R_N(1))k)} . \tag{8}$$

The discrete Fourier transform defined by (7) has computational complexity of order $O\left(N^2\right)$. Fast Fourier transform algorithms (FFT, [61], [62]) reduce the computational complexity to $O\left(N \cdot \log(N)\right)$ and are therefore typically applied instead of DFT. In classification tasks, where usually only the relatively small number M of spectral coefficients is needed, the predominance of FFT over DFT decreases, as a typical implementations of fast Fourier transform compute always all the N output coefficients. In contrast, applying the formula (7) enables to compute exactly M values, so that the actual computational complexity is of $O(M \cdot N)$ order, which may even fall below $O\left(N \cdot \log(N)\right)$ if $M < \log\left(N\right)$.

A solution to the problem of efficient Fourier descriptor computation for boundary - based shape recognition was recently proposed in [63]. This approach is based on fast adaptive Fourier transform (AFFT, [64]) which automatically chooses the sampling frequency for the analyzed contour so that the aliasing error of the obtained representation is lower than a predefined threshold. In this way the number of input points N may be significantly reduced without loss of classification effectiveness, which in turn decreases the difference between N and a fixed value of output coefficients M.

Adaptive Fourier descriptors (AFD) computed with the AFFT were successfully tested against SQUID database of sea fish contours [63]. Additionally to the original content of the database [65] several image variants obtained by rotations, horizontal and vertical flips, scaling and noise addition were included into the study in order to test the invariance of AFD resulting from the application of the normalization procedure (8).

The classification tests showed that the descriptor length M equal to 8 was sufficient to reach classification accuracy of 73.75%. Given the number of available contour points N between 400 and 1400 the process of contour descriptor computation based on fast Fourier transform needed 5.09ms and the direct formula (7) was more efficient (4.87ms). For higher M values the FFT algorithm was naturally faster than DFT but in was also observed that increasing M over the value of 16 did not improve the classification accuracy in the case of the analyzed database.

In the tests involving the AFD application the error threshold was set on a level ensuring the same classification accuracy as for FFT- and DFT-based descriptors. It was shown that in this case the mean number of necessary input points N was equal to 128, which enabled to reduce the mean time needed to compute the descriptors to 1.46ms. The tests were repeated for different numbers of output coefficients $M = 8, 16, 32, 64$ and in all the cases the mean computation time was from 3.14 to 3.74 times shorter for adaptive Fourier descriptors in comparison to the FFT-based approach (Fig. 1).

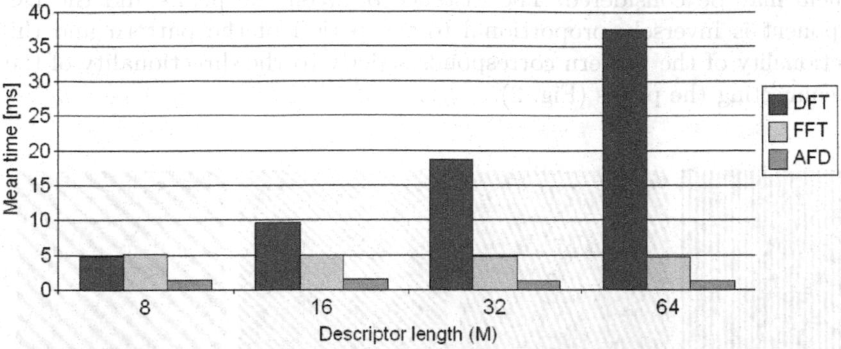
Mean time [ms]
0
5
10
15
20
25
30
35
40
8
16
32
64
Descriptor length (M)
DFT
FFT
AFD

Fig. 1 Mean times of classification for different descriptor length

The presented results confirmed the usefulness of frequency domain methods for obtaining robust shape descriptors, invariant to affine transforms. The adaptive Fourier descriptor was shown to substantially reduce the computation time in comparison to other algorithms based on Fourier transform, while retaining the same level of classification accuracy. The efficiency and effectiveness of this descriptor indicates its applicability to a wide spectrum of image classification and retrieval problems involving contour-based shape analysis and recognition.

5.3 Texture Descriptors

Periodicity and directionality of periodic patterns belong to the most important texture properties suitable for descriptor construction and classification. These characteristics, however, are difficult to be properly described by time domain methods. Frequency analysis seems to offer a more natural approach, as it inherently deals with periodic image components of known directionality. In this section we will present a method of discrete Fourier transform (DFT) application for texture descriptor construction [66].

Two-dimensional discrete Fourier transform of an $M \times N$-dimensional image f and its power spectrum are defined as [66]:

$$F(u,v) = \frac{1}{MN} \sum_{x=0}^{M-1} \sum_{y=0}^{N-1} f(x,y) e^{-j2\pi(ux/M + vy/N)} \ , \tag{9}$$

$$P(u,v) = |F(u,v)|^2 = (\text{Real}\{F(u,v)\})^2 + (\text{Imag}\{F(u,v)\})^2 \ , \tag{10}$$

where $u = 0, 1, ..., M-1$, $v = 0, 1, ..., N-1$.

Each elementary periodic pattern in the original texture is represented as a pair of distinct peaks in the power spectrum. In practice, as the power spectrum of a real image is always symmetric around the origin, only one of them may be considered. The distance between the peaks and the DC component is inversely proportional to the period of the pattern and the directionality of the pattern corresponds strictly to the directionality of the line connecting the peaks (Fig. 2).

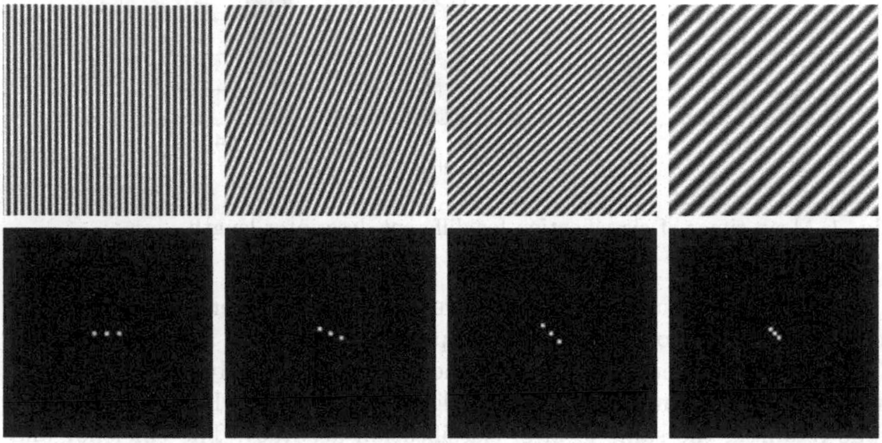

Fig. 2 Examples of four elementary patterns (top) and their power spectra (bottom). The middle peak corresponds to the DC component.

The upper half of the power spectrum (the lower half is symmetric and therefore redundant) may be expressed in polar coordinates as a function $S(r, \theta)$, where S is the power spectrum and r, θ correspond to the radial and angular coordinate respectively. If we fix $\theta = \theta_0$, the function $S(r, \theta)$ describes the spectral coefficients along the direction θ_0, which corresponds to patterns of the same directionality, having various basic frequencies. Summing $S(r, \theta_0)$ over all possible values of r yields a single number describing the overall energy of the patterns distributed along the θ_0 direction. Such a description may be computed for each θ with the following function:

$$S(\theta) = \sum_{r=1}^{R} S(r, \theta) \,, \tag{11}$$

where R is the number of valid radial coordinate values, dependent on the applied cartesian-to-polar mapping.

Similarly, fixing $r = r_0$ we obtain the coefficients located on a circle in the frequency domain, describing patterns with the same period and different directionality. These coefficients may be summed-up to form a description of all patterns having the same periodicity:

$$S(r) = \sum_{\theta=0}^{\pi} S(r, \theta) . \tag{12}$$

Both one-dimensional functions $S(\theta)$ and $S(r)$ represent the analyzed texture in a compact form enabling to asses the orientation and granularity of the dominating periodic patterns. These functions may be further described by means of some additional characteristics, e.g. the amplitude mean, variance or the global maximum position.

The presented method of frequency domain-based texture description enables to represent significant visual aspects of a texture in a compact and comprehensive form. In practical applications the efficiency of the analysis and the descriptor construction process may be easily enhanced by replacing the DFT formula (9) with a fast Fourier transform algorithm. Adaptive approach to two-dimensional fast Fourier transform construction [67] may appear even more advantageous due to the possibility of choosing the resolution level most adequate for proper description. It is also worth noting that after the periodic components have been localized in the frequency representation of the texture, they may be filtered out, and the remaining content of the texture may be further described with statistical methods, which offers a possibility of a yet more robust multi-technique texture descriptor construction.

5.4 Fourier-Mellin Transform and Its Application to Content-Based Image Retrieval

Fourier-Mellin transform (FMT) is an analytical tool applied in pattern recognition tasks for three decades now [68], [69], [70]. It is especially suited for shape description of gray-level objects as it is inherently independent on rotation, translation and scaling (the comparison between FMT and the alternative approach based on moment invariants may be found in [10]). Moreover, the generality of the representation obtained with FMT enables to perform successful content-based recognition of raw, unsegmented images, which was demonstrated by Milanese and Cherbuliez on a database containing still images extracted from video sequences [2]. Analytical Fourier-Mellin transform (AFMT) introduced in [71] was investigated by Derrode and Ghorbel in the context of image reconstruction from AFMT-based descriptors [59] and its application to shape analysis and symmetry detection was presented in [72].

The main objective of our research [73], [74] was to assess the effectiveness of FMT in the task of classification and categorization of 3D objects on the basis of their 2D representations. Two possible approaches to the problem of obtaining the invariance to translations were applied and compared on several datasets including COIL database [75].

5.4.1 Fourier-Mellin Transform

Fourier-Mellin transform (FMT) of a gray-level image represented by function f in polar coordinate system is defined as:

$$M_f(v,k) = \frac{1}{2\pi}\int_0^\infty \int_0^{2\pi} f(r,\theta)\, r^{-iv} e^{-ik\theta} d\theta \frac{dr}{r}\,, \tag{13}$$

where $(v,k) \in \Re \times Z$, $\theta \in [0, 2\pi)$, $r > 0$.

One of the ways to solve the problem of the calculus (13) convergence in the vicinity of the origin of coordinates is to compute the FMT of function $f_\sigma(r,\theta) = f(r,\theta) \cdot r^\sigma$ for a fixed value of $\sigma > 0$. The definition (13) is then transformed to the form:

$$M_{f_\sigma}(v,k) = \frac{1}{2\pi}\int_0^\infty \int_0^{2\pi} f(r,\theta)\, r^{\sigma-iv} e^{-ik\theta} d\theta \frac{dr}{r}\,, \tag{14}$$

which is referred to as analytical Fourier-Mellin transform (AFMT) [59].

Rotation and scaling invariance

It should be noted that for an image g, being a rotated and scaled variant of f the following holds true:

$$M_{g_\sigma}(v,k) = a^{-\sigma+iv} e^{ik\alpha} M_{f_\sigma}(v,k)\,, \tag{15}$$

where $g(r,\theta) = f(ar, \theta + \alpha)$.

This property enables to obtain image representation invariant to rotation and scaling. Taking the relation $M_f = M_{f_0}$ into account, we see that the FMT amplitude spectrum is indeed invariant. Hence we define:

$$I_f(v,k) = |M_f(v,k)|\,. \tag{16}$$

The AFMT spectrum needs an additional complex multiplication:

$$I_{f_\sigma}(v,k) = M_{f_\sigma}(0,0)^{\frac{-\sigma+iv}{\sigma}} e^{-ik\cdot\arg(M_{f_\sigma}(0,1))} M_{f_\sigma}(v,k)\,. \tag{17}$$

Translation invariance

The formula (15) corresponds to the shift theorem for the case of Fourier transform. If functions f and g represent two images in the Cartesian coordinate system, identical up to translation by vector $[a,b]$, i.e. $g(x,y) =$

$f(x-a, y-b)$, and F_f, F_g denote their Fourier transforms then the following holds for $\chi, \upsilon \in \Re$:

$$F_g(\chi, \upsilon) = e^{-i(a\chi + b\upsilon)} F_f(\chi, \upsilon) \ . \quad (18)$$

As the moduli of both sides of (18) are equal, we define:

$$P_f(\chi, \upsilon) = |F_f(\chi, \upsilon)| \ , \quad (19)$$

thus obtaining independence of the P_f representation from the input image translation.

5.4.2 Invariant Image Signature Construction

Image signature invariant to rotation, translation and scaling (RTS) is computed in two stages. In the first stage the invariance to translation is obtained and the remaining two transforms are considered in the second stage.

Stage 1 – Translation invariance

Applying the discrete version of formula (19) to the spectrum of the input image leads directly to shift invariance (variant I). It should be noted, however, that discarding the phase spectrum results in losing some part of shape-relevant information (Figure 3). Therefore, preliminary object positioning based on e.g. the center of mass may be applied instead [59]. In this case (variant II) the input representation for the second stage is not the amplitude of Fourier spectrum P_f but the original image after the object position adjustment.

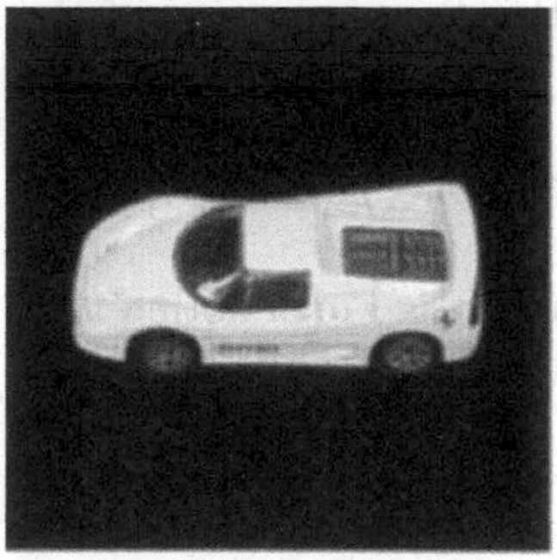

(a)

(b)

Fig. 3 Input image (a) and its reconstruction (b) based of the phase Fourier spectrum only (the whole amplitude spectrum has been set to a single constant value before the reconstruction)

Stage 2 – AFMT implementation

The discrete realization of AFMT implies changing the input coordinate system from Cartesian to log-polar [2] in order to enable application of the efficient fast Fourier transform (for comparison with other approaches see [59]). The coordinate system change is performed with the log-polar transform (LPT) according to the formula:

$$L(n,k) = P_f\left(e^{n\Delta r}\cos(k\Delta\phi) + R, e^{n\Delta r}\sin(k\Delta\phi) + R\right) , \tag{20}$$

where $n = 0, 1, ..., N-1; k = 0, 1, ..., K-1$ and

$$\Delta\phi = \frac{2\pi}{K}, \Delta r = \frac{\log(R)}{N-1}, R = \frac{A-1}{2} . \tag{21}$$

It is assumed that P_f (19) is of size $A \times A$ and it has a standard form with the DC component placed in the middle of the spectrum. If the first stage was based on the approach using the center of mass (variant II), the spectrum P_f in (20) should be replaced by the repositioned input image.

Taking the σ parameter (14) into consideration, the additional operation on the L representation must be performed:

$$L_\sigma(n,k) = e^{n\Delta r\sigma}L(n,k) , \tag{22}$$

and the fast Fourier transform (FFT) is then computed, which is equivalent to the AFMT defined in the continuous form by the formula (14):

$$M_\sigma(p,q) = \mathrm{FFT}_{2\mathrm{D}}\left\{L_\sigma(n,k)\right\} . \tag{23}$$

The spectrum M_σ, brought to the standard form with the DC component in the center (m_N, m_K), is modified with the discrete version of the equation (17):

$$I_\sigma(p,q) = M_\sigma(m_N, m_K)^{\frac{-\sigma+i(p-m_N)}{\sigma}} e^{-i(q-m_K)\arg(M_\sigma(m_N,m_K+1))} M_\sigma(p,q) , \tag{24}$$

thus obtaining the target invariant image signature. It is worth noting that the effective size of the I_σ representation is $1/2\,(N \times K)$ for the variant II based on the center of mass and $1/2\,(N \times K/2)$ for the variant using the amplitude spectrum (19), which results from the symmetry property of the real signal amplitude spectrum (Stage 1) and from the relation:

$$M_\sigma(m_N - p, m_K - q) = M_\sigma^*(m_N + p, m_K + q) . \tag{25}$$

Let us consider differences resulting from the choice of variant in the first stage. The application of the center of mass results in computing (20) for the picture in the time domain, which may lead to significant discrepancies of the LPT results (Figure 4).

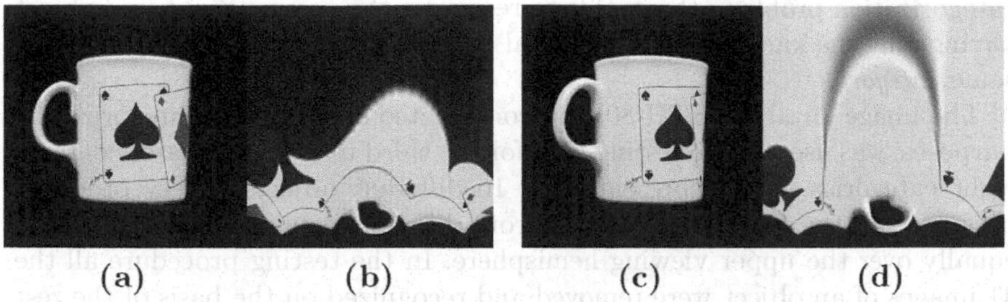
(a)
(b)
(c)
(d)

Fig. 4 Input images (a), (c) transformed with LPT (b), (d)

This observation suggests that this variant may be less suitable for recognition of 3D objects observed from different points of view, due to the difficulties in determining the origin of the coordinate system.

5.4.3 Experimental Verification

The basic material used for testing comprised three databases and independent tests were performed for each of them. All the datasets were composed of gray-level images containing single, segmented objects.

The aim of the first test was to verify the invariance of the AFMT-based descriptors to rotation, translation and scaling on the specially prepared image set. The first database (RTS-8) included eight classes, where each class contained images of one object in 24 poses obtained by planar rotation and scaling (total: 192 images). Two variants of the RTS-8 database were tested independently: RTS-8a constructed using scaling factors: 0.85, 0.9, 1, 1.07 and RTS-8b using scaling factors: 0.7, 0.8, 1, 1.14. Each scaling factor was combined with all of the rotation angle values from the set $0°, 30°, 60°, 173°, 180°, 315°$ in case of both RTS-8a and RTS-8b datasets. The preliminary tests indicated that adding translation did not alter the classification results significantly.

The second test was intended to verify classification capabilities of the AFMT for the case of 3D objects observed from different points of view. The COIL database (Columbia Object Image Library [75]) containing 20 objects, each represented by 72 images, was used for this purpose. The images were collected by putting the objects on a turntable which was then rotated by 5° for each image (for further details see [75]). For the purposes of our tests the COIL dataset was divided into training and testing dataset, each containing 36 images of all the objects.

The third testing procedure was designed to assess the applicability of the AFMT for 3D object categorization. The difference between this test and the previous one lies in the different amount of information available during recognition. In the previous test every object from the testing dataset was also present in the training data, although in different positions. In this case, in categorization problem, the goal is to recognize the previously unseen object having only the knowledge about visual appearance of other objects from the same *category*.

The image database ETH-80 [76], constructed specially for categorization purposes, was used as the testing base for the third test. This dataset contains eight categories, each represented by 10 different objects. Every object is presented in the form of 41 images from different camera locations spaced equally over the upper viewing hemisphere. In the testing procedure all the 41 images of an object were removed and recognized on the basis of the rest of the dataset.

Several signature sizes $N \times K$ (see eq. 20) were tested during the first and the second test as presented in Table 1. The presented results show that discarding the phase spectrum leads to lowering of the classification capabilities of the Fourier-Mellin image descriptor for the RTS-8b image set only. Although retaining the phase information enabled to achieve 100% recognition even of the images produced with wider range of the applied scale factors, it is clear, however, that the same does not hold true for images with other distortions introduced by varying observation point. The recognition results for COIL database were significantly better when using translation invariance based on amplitude spectrum. This problem seems to result from the ambiguity concerning the precise location of the origin of coordinates demonstrated in Figure 4.

As exactly the same issue would appear in the case of the ETH-80 database, the third test was therefore performed on the basis of variant I. Table 2 presents the recognition results and the confusion matrix. The signature size for the third test was set to 16×16.

This test presented definitely higher level of difficulties as compared to the previous ones. The results indicate high sensitivity of the investigated method to shape similarity. For example, the pears, having very distinctive shape, are recognized perfectly while the four-legged animals, especially dogs and horses, are often mixed up.

The total result of 75.64% is comparable to the mean result (80.87%) obtained by Leibe and Schiele in [76] for seven different methods including

Table 1 Classification results (test 1 and 2)

Variant	Signature size	RTS-8a	RTS-8b	COIL
I	8x8	100,00%	84.38%	99.31%
I	16x16	100,00%	91.67%	99.03%
I	32x32	100,00%	94.79%	99.17%
II	8x8	96.88%	97.40%	87.78%
II	16x16	100,00%	100.00%	93.89%
II	32x32	100,00%	100.00%	94.86%

Table 2 Classification results and the confusion matrix (test 3)

	apple	car	cow	cup	dog	horse	pear	tomato	Recognition
apple	368	0	0	5	0	0	0	37	89.75%
car	0	344	27	0	18	17	4	0	83.9%
cow	0	25	260	2	56	67	0	0	63.41%
cup	33	0	0	354	0	0	0	23	86.34%
dog	2	13	47	1	195	147	5	0	47.56%
horse	0	16	66	0	155	172	1	0	41.95%
pear	0	0	0	0	0	0	410	0	100%
tomato	32	0	0	0	0	0	0	378	92.19%

texture and color histograms, the shape context approach [52] and primary component analysis. Although generally worse by 5%, our results for a few classes, i.a. apples and pears, are indeed better than any of those obtained with methods investigated in [76] where the percentage of proper classification ranged from 57.56% to 88.29% in the case of apples and from 66.10% to 99.76% for pears.

It should be stressed that the third test alone does not fully explore the potential of the Fourier-Mellin transform, as it does not contain the element of scaling. Also the need for invariance to planar rotations does not seem to be fundamental, as both the testing set (i.e. the 41 images of an object) and the training set include the same poses of the depicted objects. Naturally, for every signature construction method there is a trade-off between the level of invariance it offers and its discriminative power [39]. However, in real-world applications the RTS invariance of FMT-based descriptors, demonstrated in the first test, is likely to be robust enough to provide high level of image recognition and retrieval also in the presence of affine transforms and other changes of visual appearance of the recognized objects.

5.4.4 Conclusion

In this section two variants of image descriptors based on Fourier-Mellin transform were investigated and applied for several image recognition and classification tasks. The presented descriptors are robust to both rigid affine transforms and other distortions introduced by changing camera location in 3D space. The difference between two variants of obtaining the translation independence was analyzed and application areas were indicated.

6 Summary

In this chapter frequency domain methods suitable for image signature construction and content-based image retrieval have been presented. The basic concepts of image recognition and classification have been introduced and the existing approaches, prototype systems, solutions and standards have been summarized. The main image feature types, both in the time and the frequency domain, as well as the current research directions have been discussed.

Three main application areas of spectral analysis: texture, contour and region-based shape recognition were discussed and the application of Fourier-Mellin transform for rotation-, translation- and scaling-invariant image signature construction was covered in detail. The results of the tests showed the usefulness of the investigated signature for content-based image classification which makes it a good candidate for image retrieval in multimedia database applications.

References

1. Subrahmanian, V.S.: Principles of Multimedia Database Systems. Morgan Kaufmann Publishers Inc., San Francisco (1998)
2. Milanese, R., Cherbuliez, M.: A rotation-, translation-, and scale-invariant approach to content based image retrieval. J. Visual Comm. Image Rep. 10, 186–196 (1999)
3. Smeulders, A.W., Worring, M., Santini, S., Gupta, A., Jain, R.: Content-based image retrieval at the end of the early years. IEEE Trans. Pattern Analysis and Machine Intelligence 22(12), 1349–1380 (2000)
4. Datta, R., Joshi, D., Li, J., Wang, J.Y.: Image Retrieval: Ideas, Influences, and Trends of the New Age. Penn State University Technical Report CSE 06–009 (2006)
5. Tadeusiewicz, R., Flasinski, M.: Pattern recognition. Polish Scientific Publishers, PWN (1991)
6. Malina, W.: The foundations of automatic image classification (in Polish). Technical University of Gdansk Press (2002)
7. Carson, C., Belongie, S., Greenspan, H., Malik, J.: Blobworld: Image segmentation using expectation-maximization and its application to image querying. IEEE Trans. Pattern Analysis and Machine Intelligence 24(8), 1026–1038 (2002)
8. Li, J., Wang, J.Z.: Automatic linguistic indexing of pictures by a statistical modeling approach. IEEE Trans. Pattern Analysis and Machine Intelligence 25(9), 1075–1088 (2003)
9. Li, J., Gray, R.M., Olshen, R.A.: Multiresolution image classification by hierarchical modeling with two dimensional hidden markov models. IEEE Trans. Information Theory 46(5), 1826–1841 (2000)
10. Grace, A.E., Spann, M.: A comparison between Fourier-Mellin descriptors and moment based features for invariant object recognition using neural networks. Pattern Recognition Letters 12, 635–643 (1991)
11. Pontil, M., Verri, A.: Support vector machines for 3d object recognition. IEEE Transactions on Pattern Analysis and Machine Intelligence 20(6), 637–646 (1998)
12. Aslandogan, Y.A., Yu, C.T.C., Liu, C., Nair, K.R.: Design, Implementation and Evaluation of SCORE. In: Proceedings of the 11th International Conference on Data Engineering (1995)

13. Marcus, S., Subrahmanian, V.S.: Foundations of multimedia database systems. Journal of ACM 43(3) (1996)
14. Li, J.Z., Özsu, M.T., Szafron, D., Oria, V.: MOQL: A multimedia object query language. In: Proceedings of the 3rd International Workshop on Multimedia Information Systems (1997)
15. Li, W.S., Candan, K.S.: SEMCOG: A Hybrid Object-based Image Database System and Its Modeling, Language and Query Processing. In: Proceedings of the 14th International Conference on Data Engineering (1998)
16. Melton, J., Eisenberg, A.: SQL Multimedia and Application Packages (SQL/MM). SIGMOD Record 30(4), 97–102 (2001)
17. ISO/IEC 13249–5:2003, Information Technology – Database Languages – SQL Multimedia and Application Packages – Part 5: Still Image (2003)
18. Oracle9i interMedia Users Guide and Reference, Release 2 (9.2). Oracle (2002)
19. Martinez, J.M.: MPEG-7 Overview, `http://www.chiariglione.org/mpeg/standards/mpeg-7/mpeg-7.htm`
20. ISO/IEC 15938–3, Information technology – Multimedia Content Description Interface: Visual
21. Flickner, M., Sawhney, H., Niblack, W., Ashley, J., Huang, Q., Dom, B., et al.: Query by image and video content: the qbic system. IEEE Computer 28(9) (1995)
22. Stricker, M., Orengo, M.: Similarity of color images. In: SPIE Storage and Retrieval for Image and Video Databases III, vol. 2185, pp. 381–392 (1995)
23. Jain, A.K.: Fundamentals of Digital Image Processing. Prentice Hall, Englewood Cliffs (1989)
24. Swain, M.J., Ballard, D.H.: Color Indexing. International J. of Computer Vision 7(1), 11–32 (1991)
25. Gong, Y., Zhang, H.J., Chua, T.C.: An image database system with content capturing and fast image indexing abilities. In: Proc. IEEE International Conference on Multimedia Computing and Systems, Boston, pp. 121–130 (1994)
26. Smith, J., Chang, S.F.: Visualseek: a fully automated content-based image query system. In: Proc. ACM Multimedia (1997)
27. Pass, G., Zabith, R.: Histogram refinement for content-based image retrieval. In: IEEE Workshop on Applications of Computer Vision, pp. 96–102 (1996)
28. Huang, J., Kumar, S.R., Mitra, M., Zhu, W.J., Zabih, R.: Image indexing using color correlogram. In: IEEE Int. Conf. on Computer Vision and Pattern Recognition, Puerto Rico, pp. 762–768 (1997)
29. Finlayson, G.D.: Color in perspective. IEEE Trans on Pattern Analysis and Machine Intelligence 8(10), 1034–1038 (1996)
30. Gevers, T., Smeulders, A.W.M.: Content-based image retrieval by viewpoint-invariant image indexing. Image and Vision Computing 17(7), 475–488 (1999)
31. Gevers, T., Smeulders, A.W.M.: Pictoseek: Combining color and shape invariant features for image retrieval. IEEE Trans. on image processing 9(1), 102–119 (2000)
32. Picard, R.W., Kabir, T., Liu, F.: Real-time recognition with the entire Brodatz texture database. In: Proc. IEEE Int. Conf. on Computer Vision and Pattern Recognition, pp. 638–639 (1993)
33. Ojala, T., Pietikainen, M., Harwood, D.: A comparative study of texture measures with classification based feature distributions. Pattern Recognition 29(1), 51–59 (1996)

34. Voorhees, H., Poggio, T.: Computing texture boundaries from images. Nature 333, 364–367 (1988)
35. Zhang, J., Tan, T.: Affine invariant classification and retrieval of texture images. Pattern Recognition 36(3), 657–664 (2003)
36. Ma, W.Y., Manjunath, B.S.: Image indexing using a texture dictionary. In: Proc. of SPIE Conf. on Image Storage and Archiving System, vol. 2606, pp. 288–298 (1995)
37. Kankanhalli, A., Zhang, H.J., Low, C.Y.: Using texture for image retrieval. In: Third Int. Conf. on Automation, Robotics and Computer Vision, pp. 935–939 (1994)
38. Schiele, B., Crowley, J.L.: Recognition without Correspondence using Multidimensional Receptive Field Histograms. International Journal of Computer Vision 36(1), 31–52 (2000)
39. Long, F., Zhang, H.J., Feng, D.D.: Fundamentals of content-based image retrieval. In: Feng, D. (ed.) Multimedia Information Retrieval and Management. Springer, Berlin (2002)
40. Ma, W.Y., Manjunath, B.S.: A comparison of wavelet features for texture annotation. In: Proc. of IEEE Int. Conf. on Image Processing, vol. II, pp. 256–259 (1995)
41. Laine, A., Fan, J.: Texture classification by wavelet packet signatures. IEEE Trans. Pattern Analysis and Machine Intelligence 15(11), 1186–1191 (1993)
42. Chang, T., Kuo, C.C.J.: Texture analysis and classification with tree-structured wavelet transform. IEEE Trans. on Image Processing 2(4), 429–441 (1993)
43. Jain, A.K., Farrokhnia, F.: Unsupervised texture segmentation using Gabor filters. Pattern Recognition 24(12), 1167–1186 (1991)
44. Tamura, H., Mori, S., Yamawaki, T.: Texture features corresponding to visual perception. IEEE Trans. on Systems, Man, and Cybernetics, Smc-8(6) (1978)
45. Liu, F., Picard, R.W.: Periodicity, directionality, and randomness: Wold features for image modeling and retrieval. IEEE Trans. on Pattern Analysis and Machine Learning 18(7) (1996)
46. Mao, J., Jain, A.K.: Texture classification and segmentation using multiresolution simultaneous autoregressive models. Pattern Recognition 25(2), 173–188 (1992)
47. Murase, H., Nayar, S.K.: Visual Learning and Recognition of 3D Objects from Appearance. International Journal of Computer Vision 14, 5–24 (1995)
48. Turk, M., Pentland, A.: Eigenfaces for Recognition. J. Cognitive Neuroscience 3, 71–86 (1991)
49. Mehrotra, R., Gary, J.E.: Similar-shape retrieval in shape data management. IEEE Computer 28(9), 57–62 (1995)
50. Berretti, S., Del Bimbo, A., Pala, P.: Retrieval by shape similarity with perceptual distance and effective indexing. IEEE Trans. Multimedia 2(4), 225–239 (2000)
51. Petrakis, E. M., Diplaros, A., Milios, E.: Matching and retrieval of distorted and occluded shapes using dynamic programming. IEEE Trans. Pattern Analysis and Machine Intelligence 24(4), 509–522 (2002)
52. Belongie, S., Malik, J., Puzicha, J.: Matching Shapes. In: International Conference on Computer Vision, ICCV 2001 (2001)
53. Hu, M.K.: Visual pattern recognition by moment invariants. IEEE Trans. Information Theory 8(2), 179–187 (1962)

54. Kass, M., Witkin, A., Terzopoulos, D.: Snakes: Active contour models. International Journal of Computer Vision 4(3), 321–331 (1988)
55. Tomczyk, A.: Image Segmentation using Adaptive Potential Active Contours. In: Computer Recognition Systems (CORES), Wroclaw, Polska. Advances in Soft Computing, pp. 148–155. Springer, Heidelberg (2007)
56. Yang, L., Algregtsen, F.: Fast computation of invariant geometric moments: A new method giving correct results. In: Proc. IEEE Int. Conf. on Image Processing, pp. 201–204 (1994)
57. Dudek, G., Tsotsos, J.K.: Shape representation and recognition from multiscale curvature. Comput. Vision Image Understanding 68(2), 170–189 (1997)
58. Mokhtarian, F., Abbasi, S., Kittler, J.: Efficient and robust retrieval by shape content through curvature scale space. In: Int Workshop on Image Databases and Multimedia Search, Amsterdam, pp. 35–42 (1996)
59. Derrode, S., Ghorbel, F.: Robust and Efficient Fourier-Mellin Transform Approximations for Gray-Level Image Reconstruction and Complete Invariant Description. Computer Vision and Image Understanding 83, 57–78 (2001)
60. Zhang, D.S., Lu, G.: A Comparative Study of Fourier Descriptors for Shape Representation and Retrieval. In: Proc. of the Fifth Asian Conference on Computer Vision (ACCV 2002), pp. 646–651 (2002)
61. Cooley, J.W., Tukey, J.W.: An algorithm for the machine calculation of complex Fourier series. Math. Comput. 19, 297–301 (1965)
62. Cooley, J.W., Lewis, P.A., Welch, P.D.: Application of the Fast Fourier Transform to Computation of Fourier Integrals, Fourier Series and Convolution Integrals. IEEE Trans. on Audio and Electroacoustics 15(2), 79–84 (1967)
63. Puchala, D., Yatsymirskyy, M.: Fast Adaptive Fourier Transform for Fourier Descriptor Based Contour Classification. In: Computer Recognition Systems 2, vol. 45. Springer, Heidelberg (2008)
64. Puchala, D., Yatsymirskyy, M.: Fast Adaptive Algorithm for Fourier Transform. In: Proc. of International Conf. on Signals and Electronic Systems, pp. 183–185 (2006)
65. `http://www.ee.surrey.ac.uk/CVSSP/imagedb/demo.html`
66. Gonzales, R.C., Woods, R.E.: Digital Image Processing, 2nd edn. Prentice Hall, Englewood Cliffs (2002)
67. Puchala, D., Yatsymirskyy, M.: Fast Adaptive Algorithm for Two - dimensional Fourier Transform, Electrical Review 10/2007 PL ISSN 0033–2097, Sigma–Not, 43–46 (2007)
68. Casasent, D., Psaltis, D.: Scale Invariant Optical Transform. Optical Engineering 15(3), 258–261 (1976)
69. Yatagay, T., Choji, K., Saito, H.: Pattern classification using optical Mellin transform and circular photodiode array. Optical Communication 38(3), 162–165 (1981)
70. Sheng, Y., Arsenault, H.H.: Experiments on pattern recognition using invariant Fourier-Mellin descriptors. J. of the Optical Society of America 3(6), 771–776 (1986)
71. Ghorbel, F.: A complete invariant description for gray-level images by the harmonic analysis approach. PatternRecog. Lett. 15, 1043–1051 (1994)
72. Derrode, S., Ghorbel, F.: Shape analysis and symmetry detection in gray-level objects using the analytical Fourier-Mellin representation. Signal Processing 84, 25–39 (2004)

73. Stasiak, B., Yatsymirskyy, M.: Comparative analysis of image descriptors based on Fourier-Mellin transform (in Polish). In: Selected Problems of Computer Science, pp. 569–577. Academic Publishing House EXIT Warsaw (2005)
74. Stasiak, B., Yatsymirskyy, M.: Application of Fourier-Mellin Transform to Categorization of 3D Objects (in Polish). In: Proc. of the 3rd Conference on Information Technologies, pp. 185–192 (2005)
75. Nene, S.A., Nayar, S.K., Murase, H.: Columbia Object Image Library (COIL-20), Tech. Report CUCS-005-96, Columbia University (1996)
76. Leibe, B., Schiele, B.: Analyzing Appearance and Contour Based Methods for Object Categorization. In: International Conference on Computer Vision and Pattern Recognition (CVPR 2003), Madison, Wisconsin (2003)

Query Relaxation in Cooperative Query Processing

Arianna D'Ulizia, Fernando Ferri, and Patrizia Grifoni

Abstract. This chapter explores new trends of query relaxation strategies that allow to implement a cooperative query processing paradigm. This new paradigm is based on the belief that the user has an idea of what he/she wants and the system has to automatically lead him/her to formulate meaningful queries by relaxing query constraints. Three kinds of query relaxation mechanisms are investigated: semantic, structural and topological query relaxation. Moreover, as similarity is an important, widely used concept in the co-operative query processing since it supports the identification of objects that are close, we intend to give an extensive overview of existing similarity definitions and methodologies for evaluating it.

1 Introduction

Query processing is the basic mechanism for accessing data in databases and information systems . Traditional query processing models were based on the belief that "the user knows what he/she wants" and consequently he/she is able to precisely formulate the query that expresses his/her needs. However, the achievement of the Internet has increased the amount of data that is stored in a variety of different kinds of databases and accessible to both expert and non-expert users. Most of the time,

Arianna D'Ulizia
Institute of Research on Population and Social Policies - National Research Council,
Via Nizza 128, 00198 Rome, Italy
e-mail: arianna.dulizia@irpps.cnr.it

Fernando Ferri
Institute of Research on Population and Social Policies - National Research Council,
Via Nizza 128, 00198 Rome, Italy
e-mail: fernando.ferri@irpps.cnr.it

Patrizia Grifoni
Institute of Research on Population and Social Policies - National Research Council,
Via Nizza 128, 00198 Rome, Italy
e-mail: patrizia.grifoni@irpps.cnr.it

D. Zakrzewska et al. (Eds.): Meth. and Support. Tech. for Data Analys., SCI 225, pp. 167–185.
springerlink.com

such users query data without a deep knowledge of how the data are structured and they may not know how to formulate meaningful queries. This situation has led current query processing models to change the traditional belief that "user knows what he/she wants" into the belief that "the user has an idea of what he/she wants" and the system has to automatically lead him/her to formulate meaningful queries by relaxing query constraints . This paradigm is captured by the cooperative query processing models. In particular, cooperative query processing aims at supporting the user by automatically modifying the query in order to better fit the real intention of the user.

Recent works dealing with information retrieval have emphasized the use of cooperative query processing techniques [Li et al., 2002; Huh et al., 2002]. Information systems and World Wide Web data collections are usually very large and heterogeneous in structure and this gives rise to the impossibility, both for expert and non-expert users, to easily formulate pinpointed queries. To overcome this problem, most information retrieval engines for the Web as well as for personal/enterprise information systems opted for the cooperative query processing paradigm for approximate retrieval. Examples of real-world advanced database applications that support this paradigm can be found in e-commerce, digital libraries and service provisioning. For instance, in an online ticket purchasing system, the user that is looking for a "concert" of a specific singer on a particular day may receive an empty answer due to the unavailability of that kind of event (concert) in the database or the imprecision in the query formulation. The challenge is to have a system that automatically modifies the search criteria and provides similar answers, apart from the exact one, that in the example is the closest available ticket.

Generally two different basic mechanisms of cooperative query processing exist in literature: *query refinement* and *query relaxation* . Query refinement *query relaxation* refers to the addition of further information into a user query from the context or user profiles so as to enable a filtering of the query answers, whereas query relaxation refers to the process of returning neighborhood or generalized information by relaxing the search condition to include additional information.

In this chapter we intend to give a comprehensive analysis and new trends of query relaxation strategies.

Query relaxation was introduced to provide fast, approximate answers to complex aggregate queries in large data warehousing environments. There are several contexts in which an exact answer may not be required and a fast, approximate answer is preferred, or the exact answer is not possible due to the unavailability of a concept involved in the query. In particular, three different reasons can lead to relax a query:

1. one or more of the query constraints as well as the query target contain data objects that are not included in the database to be searched. For instance, the query may include the data objects "Cathedral" and "City", corresponding to the query target and a query predicate respectively, which are not present in the database, so the user receives an empty answer. These data objects could be replaced by the most conceptually similar ones computed by using a semantic similarity method. For example, the objects "Cathedral" and "City" could be replaced by the data

objects "Church" and "Town", respectively, if the database contains them, as they are the most semantically similar data objects. We refer to this relaxation as *semantic query relaxation* ;

2. in structured/semi-structured databases, such as XML or graph databases, the structure of the data specified by the query does not match the structure of the data in the database. Suppose, for instance, the user is looking for "Cathedrals" in "Florence" which have the gothic style and the year of foundation lower than 1300. The information may be stored in regional tourist guide, in Florence guide, or even in a book of Art history. As the user does not know the structural relationships among the data in the database, s/he does not know where to look in order to assure an exact match between the structure of the query and the structure of the data. In this situation, cooperative query processing allows to return answers where the structure of the data approximately matches the structure specified in the query. We refer to this kind of approximation as *structural query relaxation* ;
3. in the case of geographical queries that involve topological relationships among geographical objects, there are no objects in the geographical database that satisfy the topological relationship required by the user. Suppose, for instance, the user is looking for "Cities" that are passed-through by a "River", but there are no Cities in the database that are related to a River by the pass-through relationship. This topological query constraint can be relaxed by using a topological similarity model that allows a topological relationship to be replaced by the most similar one. We refer to this case as *topological query relaxation* .

The purpose of this chapter is to investigate these three kinds of query relaxation mechanisms that allow to implement a cooperative query processing paradigm.

The need of answers that approximately match the query specified by the user, which is the basis of the cooperative query processing, requires the evaluation of similarity . As a consequence, similarity is an important, widely used concept in the cooperative query processing as it supports the identification of objects that are close, but not equal and consequently allow to identify approximate answers to a given query. In literature different definitions of similarity have been given. In this work we intend to give an extensive overview of existing similarity definitions and methodologies for evaluating it, distinguishing among semantic, structural and topological similarity.

The remaining of this chapter is structured as follows. Section 2, 3 and 4 examine similarity definitions and methodological approaches to semantic, structural and topological query relaxation, respectively, from the literature. Moreover, a critical comparison of these approaches is carried out. Section 5 concludes the chapter.

2 Semantic Query Relaxation

Providing semantic similarity metrics is an important issue in information systems as it facilitates the identification of objects that are conceptually similar. Indeed, it allows to rewrite a query that leads to unsatisfactory results by replacing the unavailable or missing concept with the most similar one, which satisfies the user request. The first approaches to semantic similarity evaluation have been based on string or keyword matching. New trends in semantic similarity assessment highlight the use of ontologies and lexical databases, containing concept definitions, frequencies, hyperonyms, synonyms, and providing a hierarchy of concepts with is-a and part-of relationships.

The most natural way to compute semantic similarity in a hierarchy is to calculate the distance between the nodes corresponding to the items being compared (the shorter the path from one node to another, the more similar they are). For example, in [Lee et al., 1989] and [Rada et al.,1989] the authors suggest that similarity in semantic networks can be thought of as involving taxonomic (Is-a) links only, to the exclusion of other link types. Distance-based measures of concept similarity assume that the domain is represented in a network, but such measures are not applicable if a collection of documents is not represented as a network. However, a well-known problem with this approach is that in real taxonomies links do not generally represent uniform distances.

For this reason in [Lin, 1998], [Resnik, 1995], [Resnik, 1999] measurement of semantic similarity in an Is-a taxonomy is based on the notion of information content. Resnik [1995, 1999] proposes algorithms that take advantage of taxonomic similarity in resolving syntactic and semantic ambiguity. For instance, in [Resnik, 1999] the author affirms that "semantic similarity represents a special case of semantic relatedness: for example, cars and gasoline would seem to be more closely related than, say, cars and bicycles, but the latter pair are certainly more similar." He also affirms that links such as Part-of can also be viewed as attributes that contribute to similarity (see also [Sussna, 1993], [Richardson et al., 1994]). In [Lin, 1998] the author investigates an information-theoretical definition of similarity that is applicable as a probabilistic model. The similarity measure is not directly stated as in earlier definitions, but is derived from a set of assumptions, in the sense that if one accepts the assumptions, the similarity measure necessarily follows. He shows how his definition can be used to measure similarity in a number of different domains. He also demonstrates that this proposal can be used to derive a measure of semantic similarity between topics in an Is-a taxonomy. He briefly discusses these different points of view in relation to particular applications or domain models.

Several authors developed semantic query relaxation mechanisms based on the information content approach proposed by Lin [Lin, 1998] and Resnik [1995, 1999]. In particular, these mechanisms rely on ontologies and lexical databases as knowledge bases of concepts that are able to capture both the syntax and semantics of concepts.

In [Maguitman et al., 2005] the authors distinguish between "tree-based similarity" and "graph-based similarity". The former relies on taxonomies (trees) and starts from the idea that the information content of a class or topic *t* (or a concept c) is defined as *-log p(t)*; that is, as the probability of a concept increases, its informativeness decreases. Therefore the more abstract a concept, the lower its information content, as also affirmed in [Ross, 1976]. The semantic similarity between two topics t_1 and t_2 is then measured as shown in equation (1), where $t_0(t_1,t_2)$ is the lowest common ancestor topic for t_1 and t_2 in the tree, and Pr[t] represents the probability that any instance is classified under topic t.

$$\sigma_s^T(t_1,t_2) = \frac{2 * logPr[t_0(t_1,t_2)]}{logPr[t_1] + logPr[t_2]} \quad (1)$$

In practice, Pr[t] is evaluated offline by dividing the number of concepts stored in the subtree, rooted at node t, by all the concepts in the tree.

The graph-based similarity generalizes the tree-based similarity measure to exploit both the hierarchical and non-hierarchical components of an ontology. The proposed graph-based semantic similarity measure was applied to the Open Directory Project ontology. The described methodology to evaluate ranking algorithms based on semantic similarity can be applied to arbitrary combinations of ranking functions stemming from text analysis.

Another semantic query relaxation mechanism that relies on the information content approach for evaluating the concept similarity in a taxonomy is proposed by Ferri et al. [Ferri et al., 2006]. The method they proposed starts from a weighted partition taxonomy of concepts defined by using WordNet 2.1 lexical database, in which the weights are the probabilities of concepts evaluated according to the frequencies of concepts in WordNet . For evaluating the semantic similarity of concepts they used the *information content similarity* (*ics*) that is essentially defined as the maximum information content shared by the concepts divided by the information content of the compared concepts. Formally, given two concepts c_1 and c_2, the semantic similarity is evaluated by means of equation (2), where *lub* is the *least upper bound* of c_1 and c_2 in the weighted taxonomy, and *p(c)* is the probability of the concept that is given by the ratio between the frequency of the concept c, estimated using noun frequencies from large text corpora, such as the *Brown Corpus of American English* [Francis et al., 1982], and the total number of observed instances of nouns in the corpus.

$$ics(c_1,c_2) = \frac{2 * logp[lub(c_1,c_2)]}{logp(c_1) + logp(c_2)} \quad (2)$$

The weighted partition taxonomy together with the *ics* are used in the query relaxation processing to identify the conceptually similar objects. For instance, suppose the user wants to express the query:

"Find all the *Provinces* which CONTAIN a *Park*"

and suppose also that the concept of *Province* is missing in the geographical database. The navigation of the weighted partition taxonomy of WordNet, shown in Figure 1, allows to determine the conceptually connected concepts to *Province*. Therefore, the evaluation of the similarities (ics) between *Province* and each of the connected concepts, that are *Region* (ics=0.82) and *Municipality* (ics=0.90), gives

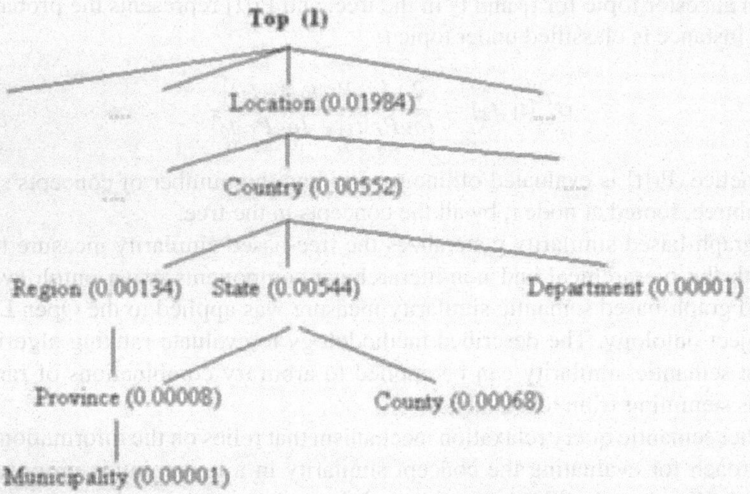
Top (1)
....
....
Location (0.01984)
....
....
Country (0.00552)
....
....
Region (0.00134)
State (0.00544)
Department (0.00001)
Province (0.00008)
County (0.00068)
Municipality (0.00001)

Fig. 1 Fragment of the weighted partition taxonomy of WordNet [Ferri et al., 2006]

as result that the query can be relaxed in one of the following two queries, ordered according to the increasing ics:

"Find all the *Municipalities* which CONTAIN a *Park*" "Find all the *Regions* which CONTAIN a *Park*"

The query relaxation mechanism proposed by Nambiar and Kambhampati [Nambiar et al., 2004] differs from the previous ones, as it doesn't rely on the information content theory, but rather on approximate functional dependencies (AFDs) [Huhtala et al., 1998]. The authors use AFDs to capture relationships between attributes of a relation and to determine a heuristic to guide the relaxation process. In particular, relaxation involves extracting tuples by identifying and executing new relaxed queries obtained by reducing the constraints on a given query. The semantic similarity between tuples is estimated as a weighted sum of similarities over distinct attributes A_i in the relation R, as shown in equation (3), where the cardinality of attributes(R) is equal to n and Wi are the weights for the attributes, whose sum is one:

$$Sim(t_1, t_2) = \sum_{i=1}^{n} Sim(t_1(A_i), t_2(A_i)) \times W_i \tag{3}$$

In the formula (3), for evaluating the similarity between two attributes Nambiar et al. [Nambiar et al., 2004] use the Jaccard coefficient [Haveliwala et al., 2002] applied to the bags of keywords associated with each attribute of the relation. As all attributes may not have the same importance for deciding the similarity among concepts, this approach, unlike the previous ones, allows to vary the weight ascribed to different attributes, giving a more realistic similarity evaluation. The main drawback of this method is the high time employed for the concept similarity estimation,

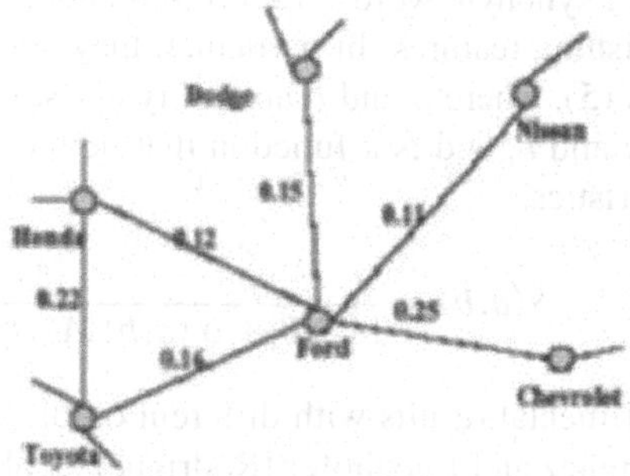
Dodge
Nissan
0.15
0.11
B M W
Honda
0.12
0.22
0.25
Ford
0.16
Chevrolet
Toyota

Fig. 2 An example of concept similarity graph in [Nambiar et al., 2004]

as it compares each concept with every other concept binding the same attribute. However, this step can be done offline, without bearing upon the query processing time. An example of concept similarity graph for the attribute *Make* of a Car is depicted in Figure 2, where the concept similarities are evaluated by using the Jaccard coefficient.

Another proposal is made in [Chien et al, 2005], where the authors examine semantic similarity between search engine queries based on their temporal correlation and develop a method to efficiently find the best-correlated queries for a given input query, which uses far less space and time. In particular they base the similarity measure on the notion of the frequency function of a query that is defined as the ratio between the number of occurrences of the query during the i_{th} time unit and the total number of queries during the i_{th} time unit. The similarity of two queries p and q is evaluated as the correlation coefficient of their frequency functions, as shown in equation (4), where $X_{j,i}$ is the frequency function, $\mu(X_j)$ is the mean frequency, and $\sigma(X_j)$ is the standard deviation of the frequency:

$$\frac{1}{d} * \sum_i \left(\frac{X_{p,i} - \mu(X_p)}{\sigma(X_p)} \right) \left(\frac{X_{q,i} - \mu(X_q)}{\sigma(X_q)} \right) \qquad (4)$$

This approach reduces the need of understanding query terms at a linguistic level, as followed by methods introduced above, and focuses on finding semantically similar queries using temporal correlation. Consequently, the similarity measure allows to identify queries whose frequencies vary greatly near a specific event, whereas it doesn't allow to represent relationships among queries that appear only a small number of times.

Besides query relaxation, semantic similarity mechanisms are well used also in the information retrieval and integration. In this field, many authors have addressed the use of semantic similarity functions for comparing objects that can be retrieved or integrated across heterogeneous databases. In particular, Rodriguez et Egenhofer [Rodriguez et al., 2003] present an approach to computing semantic similarity among entity classes of different ontologies that is based on a matching process, which takes in account both common and different characteristics between entity classes. They represent the entity classes through a description set that is composed of a set of synonym words, a set of semantic interrelations among them and a set of distinguishing features. In particular, they adopted the similarity measure given in equation (5), where a and b are entity classes, A and B corresponds to description sets of a and b, and is a function that defines the importance of the non-common characteristics.

$$S(a,b) = \frac{|A \cap B|}{|A \cap B| + \alpha(a,b)|A/B| + (1-\alpha(a,b))|B/A|} \tag{5}$$

Experimental results with different ontologies indicate that the model proposed by Rodriguez and Egenhofer [Rodriguez et al., 2003] gives good results when ontologies have complete, detailed representations of entity classes. In [Rodriguez et al., 2004] the same authors define the Matching-Distance Similarity Measure to determine semantic similarity among spatial entity classes, taking into account their distinguishing features (parts, functions, and attributes) and semantic interrelations (Is-a and Part-whole relations). Also from a spatial point of view, in [Morocho et al., 2003] the authors discuss the need for semantic integration and present a prototype information source integration tool, which focuses on schema integration of spatial databases. This tool recognizes the similarities and differences between the entities to be integrated. A domain-dependent ontology is created from the Federal Geographic Data Committee and domain-independent ontologies (Cyc and WordNet). The authors use a ratio model (ontology node distance) to assess similarities and differences between terms.

3 Structural Query Relaxation

In the field of semi-structured databases, one of the goals of the cooperative query processing is to provide a query answer such that the structure of the required data is similar to the structure specified in the query.

Most of the mechanisms of structural query relaxation rely on the representation of a query as a tree or a graph. Consequently, the structural similarity between the given query and the available data is based on how closely the structure of the query is to the structure of the data in the semi-structured database. An example of this approach is given by Barg and Wong in [Barg et al., 2003]. In their work, the authors represent a query by a tree, which indicates the required topological relationship between target nodes. For instance, to represent the query:

"Find all phone numbers of restaurants in Soho"

they use the query tree shown in Figure 3.a.

The result of the query processing is a set of answers, ranked by an overall score, which indicates how well the subgraph containing the answer satisfies the query criteria. To determine the overall score the authors use the "proximity deviation", that is the degree to which the path(n,m) satisfies the query edge criteria, where n is the node which matches the head, and m is the node matching the tail which best

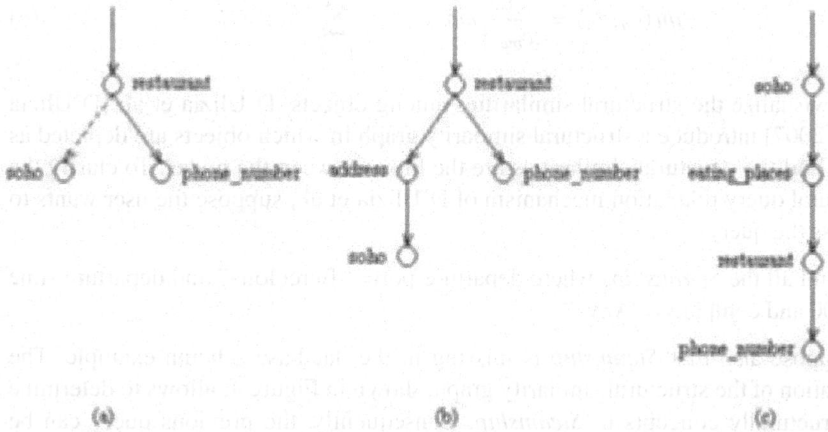
restaurant
soho
phone_number
restaurant
address
phone_number
soho
soho
eating_places
restaurant
phone_number
(a)
(b)
(c)

Fig. 3 Examples of query trees used in [Barg et al., 2003]

satisfies the query tree edge. By applying the proximity deviation scores, a possible set of relaxed queries is composed of the following queries, whose structures are represented by the trees in Figure 3.b and 3.c, respectively:

"Find all phone numbers of restaurants with an address in Soho"

"Find all phone numbers in an "eating out" guide which includes restau-rants in Soho".

The evaluation of the proximity deviation between two structures is achieved in near constant time, giving the query processing very fast performance. However, this approach is strictly devoted to query relaxation of semi-structured, in particular XML , data.

A different approach to structural query relaxation has been proposed by D'Ulizia et al. [D'Ulizia et al., 2007]. The authors don't rely on a semi-structured database, consequently they don't use the query representation as a tree or a graph, but rather they use a geographical database that contains objects, each one having a set of typed attributes. They evaluate the structural similarity between objects in a query by considering the similarity of their attributes, evaluated on the basis of the information content similarity (*ics*) . The *ics* provides the maximum information content shared by the concepts in the weighted element hierarchy of WordNet . Starting from the sets of attributes c_a and c_b of two objects, they consider the cartesian product between the two sets of attributes and they select all the sets $S_{a,b}$ of pairs of attributes such that there are no two pairs of attributes sharing an element. Then, for each of these selected sets of pairs they consider the sum of the information content similarity between the attributes of each pairs of the set. Finally, the set of pairs of attributes that maximizes the sum of the related ics is chosen and the corresponding maximal sum corresponds to the *structural similarity* between the two geographical objects. All this reasoning is expressed in the formula (6), where c_{max} is the maximal cardinality between c_a and c_b.

$$Sim(c_a, c_b) = \frac{1}{|c_{max}|} max_{S_{a,b} \in S} \sum_{\langle i,j \rangle \in S_{a,b}} ics(i, j) \tag{6}$$

To visualize the structural similarities among objects, D'Ulizia et al. [D'Ulizia et al., 2007] introduce a structural similarity graph in which objects are depicted as nodes and the structural similarities are the links between the nodes. To clarify the structural query relaxation mechanism of D'Ulizia et al., suppose the user wants to express the query:

"Find all the *Steamships* where departure-port= "Barcelona" and departure-time $\geq$10.00 and company= "Any""

Suppose also that *Steamship* is missing in the database schema example. The navigation of the structural similarity graph, shown in Figure 4, allows to determine the structurally concepts to *Steamship*. Consequently, the previous query can be transformed into the following query:

"Find all the *Airlines* where departure-air-terminal="Barcelona" and company= "Any""

Besides query relaxation, structural similarity mechanisms are well used also in the information retrieval. In this context, this involves the evaluation of distances to estimate the similarity between tree structures in terms of the hierarchical relationship of their nodes. In particular, methods for evaluating structurally similar XML documents, modeled as ordered trees, are explored. Most of these approaches for comparing semi-structured XML documents make use of techniques based on the edit distance.

[Candan et al., 2004] addresses the problem of automatically mapping semantically similar but syntactically different elements in an XML document using the inherently available structural information. In order to match two nodes in two different hierarchies, the paper proposes to map the hierarchies into a common multi-dimensional space through a data analysis method, i.e. multi-dimensional scaling, to discover the underlying spatial structure of the hierarchy's elements from the distance between them. Once the hierarchies are mapped onto the multi-dimensional

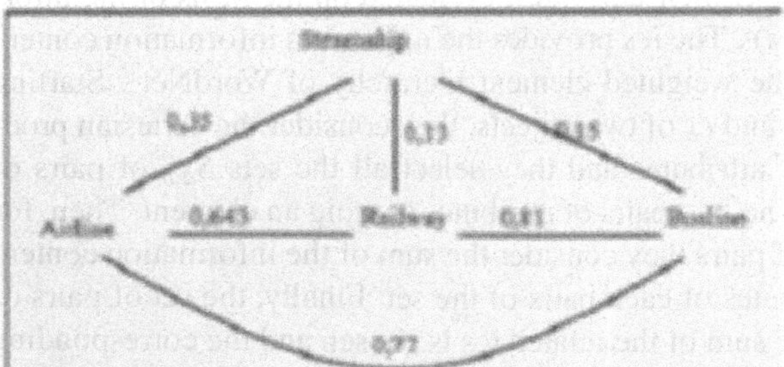
0,645
Railway
0,81
0,77

Fig. 4 The structural similarity graph in [D'Ulizia et al., 2007]

space and the transformations, required to map the common nodes from both hierarchies to each other, are identified, clustering and nearest-neighbor methods are applied to detect related uncommon nodes. As a result, the clusters, containing pairs of nodes from the two different hierarchies that are structurally similar each other, are returned. The experimental evaluation demonstrates that the approach proposed in [Candan et al., 2004] provides high degrees of precision (around 90 percent). However, this approach doesn't take into account structural repetitions/resemblances of sub-trees while comparing XML trees.

Another paper that focuses on the identification of structurally similar XML document is [Dalamagas et al., 2004]. This uses rooted ordered labeled trees [Abiteboul et al., 2000] to represent XML documents and exploits the notion of tree edit distance to estimate the structural similarity between this kind of trees, using a dynamic programming algorithm similar to Chawathe's algorithm. The main shortcoming of this edit distance algorithm consists in the application of changes to only one node at a time (using node insert, delete and update operations, with unit costs) and the consequent inaccuracy to detect sub-tree structural resemblances.

In [Tekli et al, 2007] the authors propose a method to deal with fine-grained sub-trees and leaf node repetitions, which consists of two main algorithms: one for discovering the structural commonality between sub-trees and the other for computing tree-based edit operations costs. Like the previous approach, the algorithm for discovering the structural commonality between sub-trees is based on the edit distance concept, where the maximum number of matching nodes between two sub-trees is identified with respect to the minimum number of changes (insert, delete and update operations). An experimental evaluation of this method demonstrates that it provides a high degree of precision (around 90 percent) and recall (48,7 percent), while having polynomial time complexity.

Another interesting field dealing with the computation of structural similarity is ontology matching. In this context, structural similarity refers to the measurement of distances between concepts based on the similarity of their structures in terms of attributes, types and values. [Formica et al., 2002] presents a proposal for evaluating concept similarity in an ontology management system, called SymOntos. The authors consider four notions of similarity: tentative similarity, defined as a preliminary degree of similarity with other concepts; flat structural similarity, that considers the concept's structure (attributes, values) and evaluates the similarity of each concept; hierarchical structural similarity, pertaining to hierarchically related pairs of concepts, and concept similarity, obtained by combining the structural and the tentative similarity. These notions are used to derive a method that enables the evaluation of concept similarity.

4 Topological Query Relaxation

When querying a geographical database the query is usually expressed as a set of spatial objects and a set of spatial constraints among objects. In particular, the spatial constraints can be cardinal, metric or topological constraints. Metric constraints refer to numeric values that represent the measures about areas or lengths. Cardinal constraints refer to the cardinal direction from an object to a target direction. These two kinds of constraints don't require a detailed represent numeric values and consequently they can be relaxed by varying the range in which they have to be valid. However, topological constraints are the most relevant constraints to be considered in a query on a geographical database. Before to explain how topological query constraints can be relaxed, we need to introduce some definitions.

First of all, we have to define what a topological relationship is. We consider the definition of Rodriguez et al. [Rodriguez et al., 2003], whereby "a topological relation is a binary spatial relation that is preserved under topological transformations such as rotation, translation, and scaling". Another widely used notion in the geographical query domain is the concept of configuration. According to Rodriguez et al. [Rodriguez et al., 2003], "a configuration is a set of objects and a set of constraints expressed by the binary topological relations between two objects".

In literature, most approaches to the evaluation of the similarity among topological relationships have adopted a neighborhood-based reasoning. In particular, Dylla et al. [Dylla et al, 2000] rely on a definition of topological similarity , whereby "two relations are conceptual neighbors if their spatial configurations can be continuously transformed into each other with only minimal change". Analogously, Egenhofer [1997] and D'Ulizia et al. [D'Ulizia et al., 2006] describe the similarity among topological relationships in terms of the conceptual neighborhood graph, which links most similar relations to each other.

In particular, Egenhofer in [Egenhofer, 1997] poses challenging questions with respect to the topological query relaxation and relative to the use of standard methods used in image matching and image retrieval to identify the configurations that exactly match with the geometry of the query. The author gives an answer to these questions basing the query processing on a computational model for similarity of spatial relations. In this model a query, drawn by a sketch, can be represented as a semantic network of objects and the binary relations among them. Each object corresponds to one node, while the oriented edges between nodes correspond to the binary spatial relations. Moreover, this semantic network considers five types of spatial relations: coarse and detailed topological relations, metric refining, base and detailed cardinal directions.

Focusing on coarse topological relations, Egenhofer bases the analysis of this kind of relationships on the 9-intersection model [Egenhofer, 1991], [Egenhofer and Sharma, 1993]. It represents binary topological relations between two objects, A and B, of type region, line, and point, on the basis of the nine intersections of A's interior (A ○), boundary (δA) and exterior (A-) with the interior (B○), boundary (δB) and exterior (B-) of B. The following 3x3 matrix is used to represent these criteria:

$$\begin{pmatrix} A^{\circ} \cap B^{\circ} & A^{\circ} \cap \partial B & A^{\circ} \cap B^{-} \\ \partial A \cap B^{\circ} & \partial A \cap \partial B & \partial A \cap B^{-} \\ A^{-} \cap B^{\circ} & A^{-} \cap \partial B & A^{-} \cap B^{-} \end{pmatrix}$$

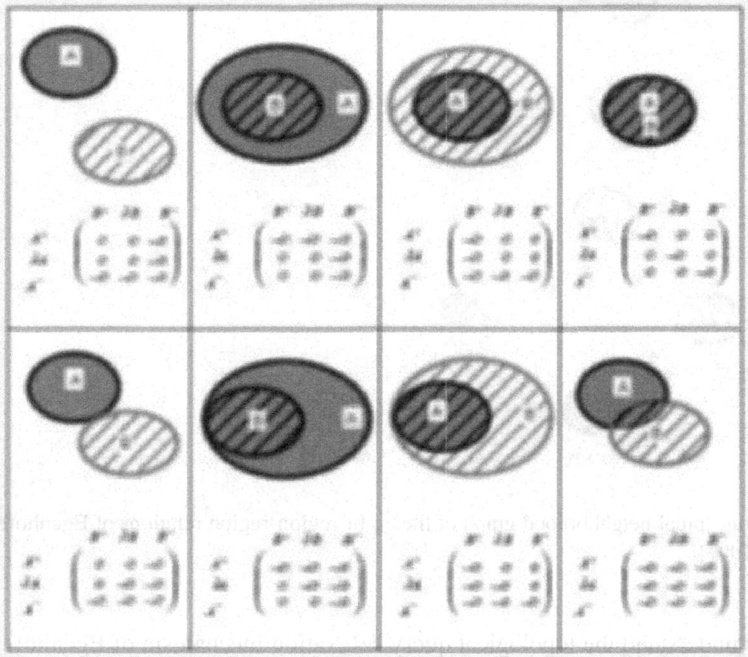

Fig. 5 The eight topological relationships between two regions represented by using the 9-intersection model of Egenhofer et al. [Egenhofer and Sharma, 1993]

The matrix values are empty (null) or not-empty (not null), depending on the intersection set (empty or not-empty). For instance, considering two regions, the topological relationships, which can be modeled by using the 9-intersection model, and their related matrices are shown in Figure 5.

To evaluate the similarity among topological relations, an order over these relations has to be introduced. To achieve that, Egenhofer proposed a conceptual neighborhood graph in which topological relations, which are the nodes of this graph, are connected if they have the least number of differences in the 9-intersection matrix. Consequently, the similarity between two topological relationships is evaluated by computing the number of differences in their 9-intersection matrices. The conceptual neighborhood graph of the eight region-region relations is shown in Figure 6.

For instance, considering the *meet* and *overlap* relations, their topological similarity is equal to 3, as the matrices differ for three elements. Egenhofer [Egenhofer, 1997] uses the conceptual neighborhood graph for relaxing a topological relation in a query by replacing it with its conceptual neighbors.

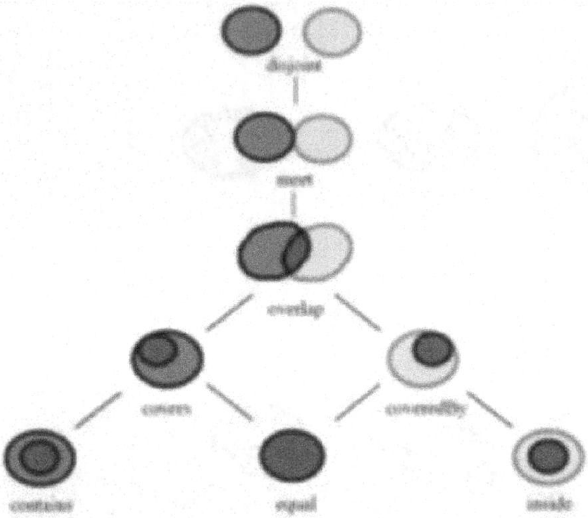

Fig. 6 The conceptual neighborhood graph of the eight region-region relations of Egenhofer [Egenhofer, 1997]

To better understand the topological query relaxation mechanism of Egenhofer, we examine an example of query. Suppose the user wants to express the query:

"Find all the *Towns* which CONTAIN a Park"

Suppose also that the spatial relations stored in the geographical database don't provide an exact match. The navigation of the conceptual neighborhood graph allows to identify the most similar topological relation to "CONTAIN", that is "COVER". Consequently the relaxed query will be:

"Find all the *Towns* which COVER a Park"

The topological query relaxation model proposed by Egenhofer is devoted to relax topological relationships between spatial objects considering a coarse set of topological relationships .

D'Ulizia et al. [D'Ulizia et al., 2006], unlike Egenhofer's approach, consider a wider set of topological configurations between spatial objects. In particular, in order to relax topological constraints, the authors define a topological similarity graph, which connects the configurations between pairs of objects, of type point, polyline or polygon, by the highest topological similarity. To represent all topological configurations between two objects, A and B, the authors introduce three different matrices, representing the point, polyline and polygon cardinalities, respectively, of the intersection of A's interior, boundary and exterior with the interior, boundary and exterior of B. These are known as the zero-dimensional 9-Intersection matrix M_0, one-dimensional 9-Intersection matrix M_1 and two-dimensional 9-Intersection matrix M_2.

$$\mathbf{M}_0 = \begin{pmatrix} |A^\circ \cap B^\circ|_0 & |A^\circ \cap \partial B|_0 & |A^\circ \cap B^-|_0 \\ |\partial A \cap B^\circ|_0 & |\partial A \cap \partial B|_0 & |\partial A \cap B^-|_0 \\ |A^- \cap B^\circ|_0 & |A^- \cap \partial B|_0 & |A^- \cap B^-|_0 \end{pmatrix}$$

$$\mathbf{M}_1 = \begin{pmatrix} |A^\circ \cap B^\circ|_1 & |A^\circ \cap \partial B|_1 & |A^\circ \cap B^-|_1 \\ |\partial A \cap B^\circ|_1 & |\partial A \cap \partial B|_1 & |\partial A \cap B^-|_1 \\ |A^- \cap B^\circ|_1 & |A^- \cap \partial B|_1 & |A^- \cap B^-|_1 \end{pmatrix}$$

$$\mathbf{M}_2 = \begin{pmatrix} |A^\circ \cap B^\circ|_2 & |A^\circ \cap \partial B|_2 & |A^\circ \cap B^-|_2 \\ |\partial A \cap B^\circ|_2 & |\partial A \cap \partial B|_2 & |\partial A \cap B^-|_2 \\ |A^- \cap B^\circ|_2 & |A^- \cap \partial B|_2 & |A^- \cap B^-|_2 \end{pmatrix}$$

For evaluating the similarity between two configurations D'Ulizia et al. compute the sums of the differences of the corresponding elements in each of the three matrices, as illustrated in the formula (7).

$$Sim(A,B) = \sum_{k=0}^{2}\sum_{i=0}^{2}\sum_{j=0}^{2} M_K^A[i,j] - M_K^B[i,j] \qquad (7)$$

The similarities values appear as labels of the edges of the topological similarity graph, while the nodes are the topological configurations between objects. For instance, considering a polyline and a polygon, the topological similarity graph is shown in Figure 7.

To better understand the topological query relaxation mechanism of D'Ulizia et al., we consider the geographical query depicted in Figure 8.a. that correspond to the following query in natural language:

"Find all the *Towns* which are PASSED-THROUGH by a *Railway*".

Suppose that there are no towns in the geographical database, which are passed through by a Railway, so the exact answer is not available. The topological similarity graph is used to find the most similar configurations to the one drawn by the user. All configurations between a polygon (Town) and a polyline (Railway) that are adjacent to the pass-through configuration are taken in consideration (Figure 8.b), and they are ordered by increasing topological similarity.

Another paper adopting a neighborhood-based reasoning is [Dylla et al, 2005]. The authors present an idea on how results of qualitative topological reasoning can be exploited when considering action and change. They investigated how conceptual neighborhood structure could be applied in the situation calculus for qualitative reasoning on relative positional information. In particular, starting from the assumption that a movement can be modeled qualitatively as a sequence of neighboring spatial

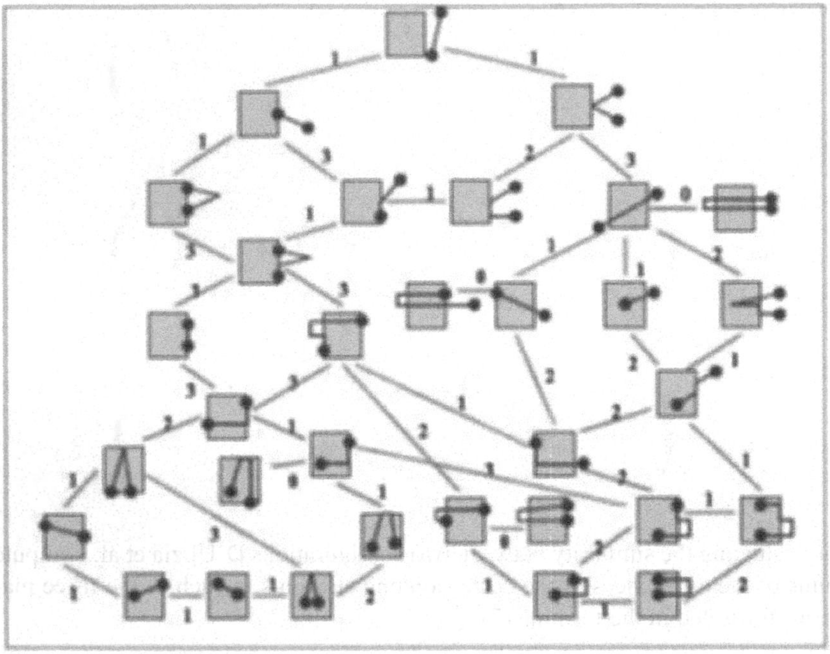

Fig. 7 The topological similarity graph for the polyline-polygon configurations of D'Ulizia et al. [D'Ulizia et al., 2006]

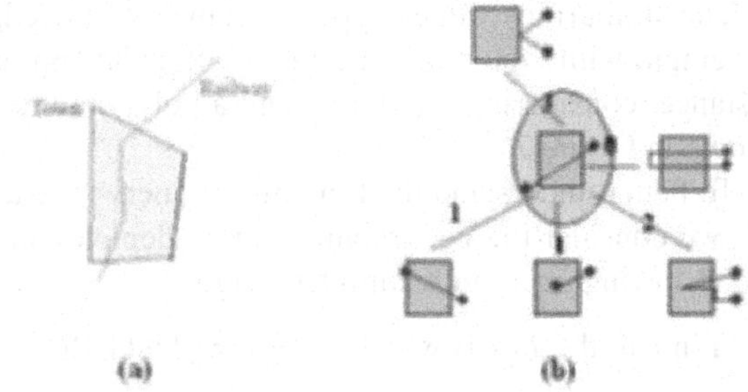
(a)
(b)

Fig. 8 An example of query and a fragment of the topological similarity graph focusing on the pass-through configuration

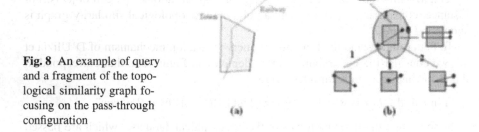

relations which hold for adjacent time intervals, the approach of Dylla and Moratz relies on dipole relation algebra for representing both spatial objects with intrinsic orientation as dipoles, and their relationships as atomic relations of the dipole calculus. This approach is strictly devoted to dynamic environment for reasoning about relative positions of spatial moving objects.

5 Conclusion

In this chapter, we explored new trends in cooperative query processing, focusing on query relaxation that allows to provide fast, approximate answers in the case an exact answer is not required or possible due to the unavailability of a concept involved in the query. According to the main reasons that lead to relax a query, we classified the query relaxation mechanisms as *semantic*, *structural* and *topological* query relaxation. The first concerns the approximation of a query that leads to unsatisfactory results by replacing the unavailable or missing concept with the most semantically similar one, which satisfies the user request. The second allows to return answers where the structure of the data approximately matches the structure specified in the query. Finally, the third concerns the approximation of a geographical query by replacing the unavailable topological relationships with the most topologically similar ones.

The notion of similarity is closely related to the query relaxation. In fact, it gives a measure of how far the data in the query specified by the user "deviates" from the data available in the database. In this chapter, we investigated existing similarity definitions and methodologies for evaluating it, according to the three kinds of query relaxation we have identified.

References

1. Abiteboul, S., Buneman, P., Suciu, D.: Data on the Web. Morgan Kaufmann, San Francisco (2000)
2. Barg, M., Wong, R.K.: Cooperative Query Answering for Semistructured Data. In: Schewe, K.-D., Zhou, X. (eds.) Proc. Fourteenth Australasian Database Conference (ADC 2003) CRPIT, Adelaide, Australia, vol. 17, pp. 209–215. ACS (2003)
3. Candan, K.S., Kim, J.W., Liu, H., Suvarna, R.: Structure-based mining of hierarchical media data, metadata, and ontologies. In: 5th Workshop on Multimedia Data Mining, Seattle, WA (2004)
4. Chien, S., Immorlica, N.: Semantic Similarity between Search Engine Queries using Temporal Correlation. In: International World Wide Web Conference Committee (IW3C2), WWW 2005, Chiba, Japan, May 10-14, pp. 2–11 (2005)
5. Dalamagas, T., Cheng, T., Winkel, K.J., Sellis, T.K.: Clustering XML Documents by Structure. In: Vouros, G.A., Panayiotopoulos, T. (eds.) SETN 2004. LNCS, vol. 3025, pp. 112–121. Springer, Heidelberg (2004)
6. D'Ulizia, A., Ferri, F., Grifoni, P., Rafanelli, M.: Relaxing constraints on GeoPQL operators for improving query answering. In: Bressan, S., Küng, J., Wagner, R. (eds.) DEXA 2006. LNCS, vol. 4080, pp. 728–737. Springer, Heidelberg (2006)
7. D'Ulizia, A., Ferri, F., Grifoni, P., Rafanelli, M.: Structural similarity in geographical queries to improve query answering. In: Proceedings of the 2007 ACM Symposium on Applied Computing (SAC), Seoul, Korea, pp. 19–23 (2007)
8. Dylla, F., Moratz, R.: Exploiting qualitative spatial neighborhoods in the situation calculus. In: Freksa, C., Knauff, M., Krieg-Brückner, B., Nebel, B., Barkowsky, T. (eds.) Spatial Cognition IV. LNCS, vol. 3343, pp. 304–322. Springer, Heidelberg (2005)
9. Egenhofer, M.J.: Query Processing in Spatial-Query-by-Sketch. Journal of Visual Languages and Computing 8(4), 403–424 (1997)
10. Egenhofer, M.J., Sharma, J.: Topological relations between regions in R2 and Z2. In: Abel, D.J., Ooi, B.-C. (eds.) SSD 1993. LNCS, vol. 692, pp. 316–336. Springer, Heidelberg (1993)

11. Egenhofer, M.J.: Reasoning about binary topological relations. In: Günther, O., Schek, H.-J. (eds.) SSD 1991. LNCS, vol. 525, pp. 143–160. Springer, Heidelberg (1991)
12. Ferri, F., Formica, A., Grifoni, P., Rafanelli, M.: Query approximation by semantic similarity in GeoPQL. In: 2nd International Workshop on Semantic-based Geographical Information Systems (SeBGIS 2006), Montpellier, France (2006)
13. Formica, A., Missikoff, M.: Concept Similarity in SymOntos: an Enterprise Ontology Management Tool. The Computer Journal 45(6), 583–594 (2002)
14. Francis, W.N., Kucera, H.: Frequency Analysis of English Usage. Houghton Mifflin, Boston (1982)
15. Haveliwala, T., Gionis, A., Klein, D., Indyk, P.: Evaluating strategies for similarity search on the web. In: Proceedings of WWW, Hawai, USA (2002)
16. Huh, S., Moon, K., Ahn, J.: Cooperative query processing via knowledge abstraction and query relaxation. In: Siau, K. (ed.) Advanced Topics in Database Research, vol. 1, pp. 211–228. IGI Publishing, Hershey (2002)
17. Huhtala, Y., Krkkinen, J., Porkka, P., Toivonen, H.: Efficient discovery of functional and ap-proximate dependencies using partitions. In: Proceedings of ICDE (1998)
18. Lee, J.H., Kim, M.H., Lee, Y.J.: Information retrieval based on conceptual distance in is-a hierarchies. Journal of Documentation 49(2), 188–207 (1989)
19. Li, W.S., Candan, K.S., Vu, Q., Agrawal, D.: Query Relaxation by Structure and Semantics for Retrieval of Logical Web Documents. IEEE Transactions on Knowledge and Data Engineering 14(4), 768–791 (2002)
20. Lin, D.: An Information-Theoretic Definition of Similarity. In: Proceedings of the 15th Intern. Conference on Machine Learning, ICML 1998, Madison, WI, pp. 296–304 (1998)
21. Maguitman, A.G., Menczer, F., Roinestad, H., Vespignani, A.: Algorithmic detection of semantic similarity. In: International World Wide Web Conference Committee (IW3C2), WWW 2005, Chiba, Japan, May 10-14, pp. 107–116 (2005)
22. Morocho, V., Perez-Vidal, L., Saltor, F.: Semantic integration on spatial databases SIT-SD prototype. In: Proceedings Of VIII Jornadas de Ingenieria del Software y Bases de Datos, Alicante, Spain, pp. 603–612 (2003)
23. Nambiar, U., Kambhampati, S.: Mining Approximate Functional Dependencies and Concept Similarities to Answer Imprecise Queries. In: Proceedings of the Seventh International Work-shop on the Web and Databases (WebDB 2004), Paris, France, pp. 73–78 (2004)
24. Rada, R., Mili, H., Bicknell, E., Blettner, M.: Development and application of a metric on semantic nets. IEEE Transaction on Systems, Man, and Cybernetics 19(1), 17–30 (1989)
25. Resnik, P.: Semantic similarity in a taxonomy: an information-based measure and its application to problems of ambiguity in natural language. Journal of Artificial Intelligence Research 11, 95–130 (1999)
26. Resnik, P.: Using information content to evaluate semantic similarity in a taxonomy. In: Proceedings of the IJCAI (1995)
27. Richardson, R., Smeaton, A.F., Murphy, J.: Using WordNet as a knowledge base for measuring semantic similarity between words. In: Working paper CA-1294, Dublin City University, School of Computer Applications, Dublin, Ireland (1994)
28. Rodriguez, M.A., Egenhofer, M.J., Blaser, A.D.: Query pre-processing of topological constraints: comparing a composition-based with neighborhood-based approach. In: Hadzilacos, T., Manolopoulos, Y., Roddick, J., Theodoridis, Y. (eds.) SSTD 2003. LNCS, vol. 2750, pp. 362–379. Springer, Heidelberg (2003)
29. Rodriguez, M.A., Egenhofer, M.J.: Comparing Geospatial Entity Classes: an Asymmetric and Content-Dependent Similarity Measure. International Journal of Geographical Information Science 18(3), 229–256 (2004)

30. Rodriguez, M.A., Egenhofer, M.J.: Determining Semantic Similarity among Entity Classes from Different Ontologies. IEEE Transactions on Knowledge and Data Engineering 15(2), 442–456 (2003)
31. Ross, S.: A First Course in Probability. Macmillan, Basingstoke (1976)
32. Sussna, M.: Word sense disambiguation for free-text indexing using a massive semantic network. In: Proceed. Second Intern. Conference on Information and Knowledge Management (CIKM 1993), Arlington, Virginia (1993)
33. Tekli, J., Chbeir, R., Yetongnon, K.: A Fine-Grained XML Structural Comparison Approach. In: Parent, C., Schewe, K.-D., Storey, V.C., Thalheim, B. (eds.) ER 2007. LNCS, vol. 4801, pp. 582–598. Springer, Heidelberg (2007)

Ensuring Mobile Databases Interoperability in Ad Hoc Configurable Environments: A Plug-and-Play Approach

Angelo Brayner, José de Aguiar Moraes Filho, Maristela Holanda, Eriko Werbet, and Sergio Fialho

Abstract. The topology of a mobile ad hoc network (MANET) may change randomly and rapidly at unpredictable time, since nodes are free to move arbitrarily. In such an environment, we may have a collection of autonomous, distributed, heterogeneous and mobile databases, denoted Mobile Database Community or MDBC. MDBC participants may join to the community as they move within communication range of one or more hosts members of the MDBC. Participants may transiently disconnect from the network due to communication disruptions or to save power. Thus, an MDBC can be characterized as an ad hoc (dynamically) configurable environment. This chapter describes an agent-based middleware, denoted AMDB

Angelo Brayner
University of Fortaleza (UNIFOR), Av. Washington Soares, 1321 CEP 60811-905 Fortaleza, CE, Brazil
e-mail: brayner@unifor.br

Jose de Aguiar Moraes Filho
University of Fortaleza (UNIFOR), Av. Washington Soares, 1321 CEP 60811-905 Fortaleza, CE, Brazil
e-mail: jaguiar@unifor.br

Maristela Holanda
Federal University of Rio Grande do Norte (UFRN), Caixa Postal 1524 – Campus Universitrio Lagoa Nova CEP 59072-970 Natal, RN, Brazil
e-mail: mholanda@dca.ufrn.br

Eriko Werbet
University of Fortaleza (UNIFOR), Av. Washington Soares, 1321 CEP 60811-905 Fortaleza, CE, Brazil
e-mail: eriko@unifor.br

Sergio Fialho
Federal University of Rio Grande do Norte (UFRN), Caixa Postal 1524 – Campus Universitrio Lagoa Nova CEP 59072-970 Natal, RN, Brazil
e-mail: fialho@pop-rn.rnp.br

D. Zakrzewska et al. (Eds.): Meth. and Support. Tech. for Data Analys., SCI 225, pp. 187–217.
springerlink.com

(Accessing Mobile Databases), which enables such communities to be formed opportunistically and in a plug-and-play manner over mobile database in ad hoc configurable environments. AMDB has a fully distributed architecture and has the capability of exploiting physical mobility of hosts and logical mobility of database queries (transactions) across mobile hosts. Furthermore, this chapter describes a query engine for the proposed architecture and a mobile transaction processing model as well.

1 Introduction

Advances in portable computing devices and wireless network technology have led to the development of a new computing paradigm, called mobile computing. This new paradigm enables users carrying portable devices to access services provided through a wireless communication infrastructure, regardless of their physical location or movement patterns. The database technology has been impacted by the mobile computing paradigm. For example, database systems may reside on mobile computers which are nodes of a mobile ad hoc network (MANET [13]). Since a MANET has a topology which may change randomly and rapidly at unpredictable time, a dynamic collection of autonomous, heterogeneous and mobile databases (denoted Mobile Database Community or MDBC) can be opportunistically formed [4]. An MDBC models a multidatabase [8, 14] whose members (local databases) can move across different locations.

Sharing information among multiple heterogeneous, autonomous, distributed and mobile data sources in MDBCs has emerged as a strategic requirement for several applications [6, 16, 18, 19, 20]. We envision a number of mobility scenarios and applications that may require mobile database sharing in MDBCs. For example, in battlefields, we may have several units of a coalition force (from different countries) moving around different geographical regions, but having a common goal. The commander of each unit has access to local database system (residing in mobile computer installed in a tank or humvee) containing information about his/her own unit and some information about the enemy forces. Thus, the databases of the multiple units of the coalition force are distributed, autonomous and mobile. They might be heterogeneous as well. In order to take tactical decisions, each commander may have to access the databases of others units of the coalition to get information about them (their strength w.r.t. type and number of weapons and number of components). For such a class of applications, sharing mobile databases belonging to an MDBC has become a critical issue.

However, sharing databases in an MDBC requires that the following problems should be addressed: (*i*) heterogeneous databases integration; (*ii*) query processing over a variable number of mobile databases and; (*iii*) mobile transaction processing (i.e., concurrency control and recovery for mobile and distributed transactions).

In this chapter we propose a middleware, denoted Accessing Mobile Database (AMDB), for supporting database sharing in mobile database communities. The key goal of the proposed middleware is to provide access to heterogeneous, autonomous and mobile databases. Therefore, AMDB provides full database system functionalities. AMDB is based on the concept of mobile agents and is fully distributed. Furthermore, AMDB provides the necessary support for forming MDBCs opportunistically over a collection of mobile databases hosts.

In order to support queries over mobile databases of an MDBC a query engine for AMDB is described. Since the query engine may operate in an unpredictable and dynamically changeable environment (MDBC), the proposed query processing strategy is adaptive to the query execution context. Finally, this chapter presents the Intelligent Transaction Scheduler (ITS), a scheduler for controlling in the AMDB middleware, which can be characterized by having a hybrid behavior (conservative or aggressive). ITS has the ability for automatically identifying the modifications in the computational environment in which it is inserted and for adapting its behavior without human interference. For that reason, ITS is a context-aware and self-adaptable scheduler. Some experiments are shown to prove that ITS is quite efficient for synchronizing transactions in a dynamically configurable environment, such as an MDBC.

It is important to note that AMDB does not require any change to the core of underlying databases systems. This property enables commercial DBMSs for participating in an MDBCs through AMDB. For that reason, we classify the AMDB approach as non-intrusive.

This chapter is structured as follows. In Section 2, a mobile computing environment is characterized. In Section 3, the concept of mobile databases community is defined and proposed architecture is presented and analyzed. Section 4 presents a query engine for processing queries submitted to AMDB, as well as a cost model. Mobile transaction processing issues in the MDBC context are addressed in Section 5. Section 6 concludes the chapter.

2 Mobile Computing Environment Model

An abstract model for a mobile computing environment can be defined as follows. Mobile computers are grouped into components called cells, where cells represent geographical regions covered by a wireless communication infrastructure. Such cells can represent a Wireless Local Area Network (WLAN), an ad hoc network, an geographical area covered by a cell phone network (called cell as well) or a combination of those communication technologies (for example, an ad hoc network inside cell of a cellular phone network).

Mobile support stations (or base stations) are components of a mobile computing platform which have a wireless interface in order to enable communication between mobile units located at different cells. Thus, each cell must be associated with one mobile support station. Communication between two base stations is made through a fixed network. In fact, mobile support stations represent fixed hosts interconnect through a high-speed wired network. From a database technology standpoint, there are two classes of database systems in a mobile computing environment: *(i)* a class consisting of database systems which reside on fixed hosts, and; *(ii)* a class of database systems which reside on mobile computers. Database systems residing on mobile hosts are denoted mobile databases.

It is worthwhile to note that, when a cell in a mobile computing environment represents a WLAN or an ad hoc network, mobile computers inside the cell can communicate with each other directly. On the other hand, when a cell represents an area covered by a cell phone network, mobile units need should use the cell's base station to communicate with each other. A mobile computer can use different communication technologies. We call vertical handoff the fact of a mobile computer migrating from a given communication technology to another. For example a mobile computer A can use a to communicate with other mobile computer inside the same cell, and it can use a cellular phone network to communicate with mobile computers located at another cell.

From a database technology perspective, we can categorize mobility in two different types:

(i) *Physical Mobility:* this type of mobility cope with spatial mobility of database servers through different space regions;
(ii) *Logical Mobility:* this type of mobility occurs in the logical space. For that reason is related to database access code (e.g. SQL expressions, stored procedures or methods) migration among several mobile database servers.

A Mobile Database Community (MDBC) is defined as a dynamic collection of distributed, autonomous, heterogeneous and mobile databases, interconnected through a wireless communication infrastructure of a mobile ad hoc network. An MDBC models a dynamically configurable multidatabase whose members (local databases in mobile hosts) can move across different locations.

A varying number of mobile computers can participate in an MDBC. New participants may join to an MDBC as they move within communication range of one or more hosts which are members of the MDBC. On other hand, MDBC participants may transiently disconnect from the network due to communication disruptions or to save power. Therefore, an MDBC represents a dynamically configurable environment.

A key characteristic of mobile databases belonging to an MDBC is that they were created independently and in an uncoordinated way without considering the possibility that they might be somehow integrated in the future.

3 Sharing Mobile Databases in MDBCs

The key goal of the proposed middleware (called Accessing Mobile Databases or AMDB, for short) is to enable mobile database communities (MDBCs) to be formed opportunistically over mobile databases residing on mobile hosts which are nodes of a MANET and to make mobile databases belonging to the formed MDBC interoperable. The architecture is fully distributed and relies on existing mobile agent technology [12, 20].

The AMDB architecture is composed by two classes of agents: stationary and mobile agents. The class of stationary agents is composed by the manager and wrapper agents. Manager agents are responsible for managing local computational resources of a mobile computer. Wrapper agents provide an interface between mobile unit users (and applications) and the AMDB platform. For that reason, wrapper agents have the following functionalities. First, they should create the execution context for mobile agents. Thus, a wrapper agent behaves like a middleware between a mobile agent engine [12] and a database system belonging to an MDBC. Second, a wrapper agent provides a common data representation (conceptual schema) of the local data source. XML will be used as a common data representation [23]. Additionally, a wrapper agent is also responsible for temporarily transferring one or more services to another mobile computer of the community to improve performance or whenever a critical situation occurs (for example, temporary service unavailability such as insufficient local memory for processing a query).

Agents of the mobile-agent class are responsible for implementing logical mobility. Consequently, they should be able to migrate from one member to another in an MDBC in order to carry database queries and results. There are three types of mobile agents: Runner, Loader and Broker agents. Runner agents are responsible for performing tasks (e.g., processing queries) required by mobile unit users in remote mobile databases. Loader agents have the functionality of carrying a database query result back to the mobile unit that has submitted the query. Broker agents should gather schemas of mobile databases of an MDBC, when a mobile unit joins to the MDBC.

Fig. 1 depicts an abstract model of the AMDB architecture. In order to illustrate how the proposed architecture works, consider that a user in mobile host MU_K wants to form an MDBC. First of all, the mobile host MU_K declares its intention to the wrapper agent by sending a *create_mdbc* command. The wrapper agent creates a broker agent and gives to it the task to negotiate the creation of an MDBC with mobile hosts inside the area covered by the MANET in which MU_K is connected. Suppose now that a mobile unit MU_I decides to join to the MDBC. The wrapper agent of MU_I creates a broker agent in order to query local database schemas (or part of it) of all databases which decide to share their databases by means of the new MDBC. Local schemas visible to the members of an MDBC are specified as XML documents.

With the database schemas of each mobile database belonging to the MDBC, users or applications are able to access the mobile databases through an interface provided by the local wrapper agent. Such queries should be specified in XQuery [24]. The idea is to use the XQuery as a multidatabase language [8, 14] to integrate mobile databases in an MDBC.

Consider that a user in MU_I submits a mobile query (a query over several mobile databases). In that case, the wrapper agent of MU_I creates a runner agent, which has to migrate to MU_K, since a mobile database belonging to the MDBC resides in MU_K. When the runner agent arrives at MU_K, it sends the XQuery expression to the local wrapper agent. The wrapper agent of MU_K, in turn, maps the XQuery expression to the native query language of the local database and submits it for execution.

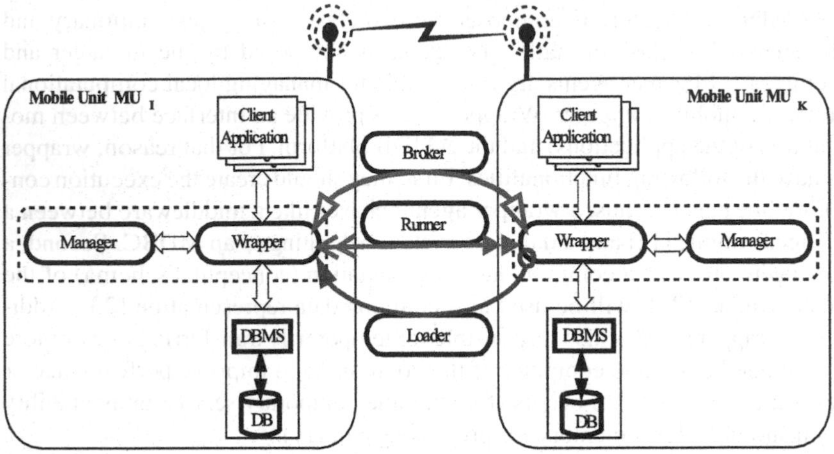
Mobile Unit MU I
Client Application
Broker
Runner
Manager
Wrapper
DBMS
DB
Loader
Mobile Unit MU K
Client Application
Wrapper
Manager
DBMS
DB

Fig. 1 The AMDB Architecture

The connectivity between a wrapper agent and a local database system is realized through a JDBC API. When the runner agent returns home, it communicates to the wrapper agent of MU_I that its task has been finished. The query engine in the AMDB architecture is fully distributed, since query processing activities are executed by different components (or agents) in different mobile units (sites). The next section the query processing mechanism for AMDB is described in details.

4 Processing Queries over Mobile Databases

Since the query engine for AMDB may operate in an unpredictable and dynamically changeable environment (MDBC), the proposed query processing strategy is adaptive to the query execution context. For that reason, besides the AMDB's query engine, an adaptive operator, denoted AJAX, and a strategy for processing queries with unavailable data sources are presented in this section.

4.1 Query Engine

As already said, users or application programs submit queries to the AMDB architecture as an XQuery expression. Such queries can be categorized in two classes: local or mobile (distributed) queries. A local query only involves operations over objects of the local database. In turn, a mobile query may involve operations over several mobile databases of an MDBC.

Fig. 2 depicts an abstract model of the query engine for processing mobile (distributed) queries in the AMDB architecture. According to the proposed engine, a query is processed by six distinct modules: Pre-Processing, Decomposition, Context Recognition, Clone Generation, Fragment Execution and Post-Processing

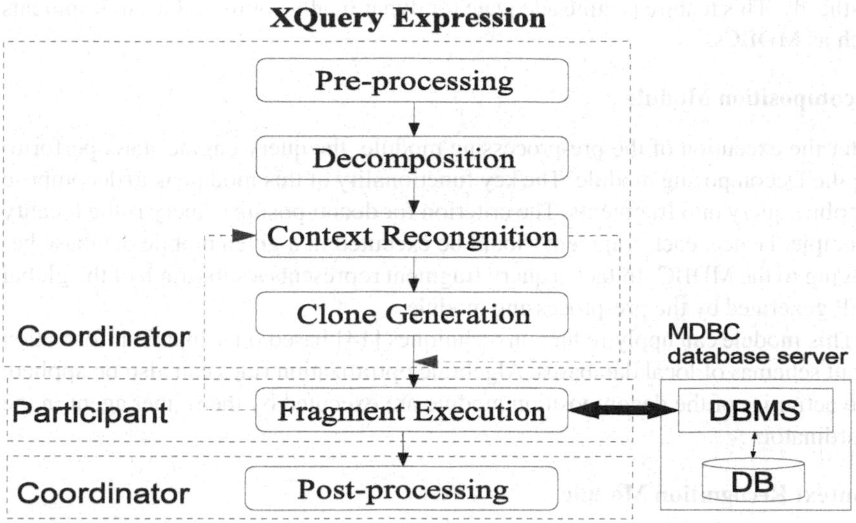
XQuery Expression
Pre-processing
Decomposition
Context Recongnition
Clone Generation
Coordinator
MDBC
database server
Participant
Fragment Execution
DBMS
DB
Coordinator
Post-processing

Fig. 2 Query Engine

modules. In fact, each module comprises activities of a phase in the query processing workflow. We denote Coordinator the origin unit (host) in which the query was initially submitted. The functionalities of the coordinator are executed by the runner, loader and wrapper agents. A Participant represents a DBS, which is responsible for executing locally operations of a global query.

Pre-processing Module

The activities of the pre-processing module are executed by wrapper and runner agents. Initially, the query submitted in an XQuery format is parsed by the wrapper agent. After that, the wrapper agent sends the parsed query to the runner agent. After receiving the query, the runner generates a query execution plan (QEP). The generated QEP is represented by an operator graph, whose nodes represent XQuery algebraic operators or data sources. The edges represent the data flow between nodes. The nodes have attributes like data locality, selectivity factor and name of data source.

In a QEP, each operator node can also be annotated with triggers which should execute actions when specified events happen in order to adapt the initial QEP to a new query execution context. Inaccurate statistics, memory overflow (e.g., when executing join operations), and unavailable data source are some of the events which could be specified. The correspondent actions could be: reformulate plan, execute a new operator (which consumes less memory) or reformulate route. Therefore, the benefit of using the concept of triggers is to provide the query engine the reacting capacity to environment changes. By doing this, the proposed query engine supports adaptive query processing, since it has the capability of changing the QEP on-the-fly. This feature is quite adequate for dynamically configurable environments such as MDBCs.

Decomposition Module

After the execution of the pre-processing module, the query engine starts performing the Decomposing module. The key functionality of this module is to decompose a mobile query into fragments. The criterion for decomposing a query is the locality principle. Hence, each fragment should be executed in a given mobile database belonging to the MDBC. In fact, a query fragment represents a sub-graph of the global QEP generated by the pre-processing module.

This module can apply reduction techniques [14] based on semantic information about schemas of local databases. Algebraic optimization rules can also be applied. The activities of the decomposition module are executed by the runner agent in the coordinator.

Context Recognition Module

Once a query has been divided into fragments, the query engine has to verify, for each fragment, whether or not mobile databases (which are nodes in the QEP) have computational resources to execute the fragments.

If a given mobile database does not have support to execute a given operation (for example, a join), the query engine itself is responsible for executing that operation. In fact, this should be performed by the runner agent and can be carried out either on the coordinator (origin unit) or on another mobile unit. In this case, it is necessary a QEP reordering, since the runner agent is now responsible for executing the operation.

After the process described above, the context recognition module applies the cost model defined in Section 4.2. Based on the result of the cost model, the engine can decide for a parallel or a linear execution of the QEP. We define a linear QEP execution (or linear QEP, for short) as a serial execution of query fragments. In this case, the clone generation module is not started, and the query fragments are sent directly to the fragment execution module (flow represented by the dashed line in Fig. 2). On the other hand, a Parallel QEP execution (or parallel QEP, for short) is defined as a parallel and asynchronous execution of fragments of a given query. In this case, the clone generation module should be activated.

An execution route represents a sequence of mobile units in which fragments of a given query should be executed. The concept of route is necessary for defining migration plans for runner agents, whenever executing a linear QEP. The key idea behind the concept of route is the following: the runner agent should visit each mobile unit specified in the route in an ordered way, that is, the first mobile unit in route is visited firstly; the second mobile unit is visited secondly, and so on. The route for a liner QEP is defined based on the heuristic H1, presented in Table 1.

The decision of executing a given QEP as a linear or parallel QEP is based on heuristic H2 (see Table 1).

Table 1 Heuristics

Heuristic1 (or Route Heuristic): The route is defined according to the cardinality of the intermediate results produced by each mobile database. The sequence should reflect an ordering of such cardinalities (the list begins with the mobile unit with lowest cardinality, following the mobile unit with the second lowest cardinality, and so on).	**(H1)**
Heuristic2: A QEP should be executed in parallel, if there exists no statistics at all or the cost of the parallel QEP is less than the cost of the linear QEP (defined by the route heuristic).	**(H2)**

Clone Generation Module

Once the execution of the context recognition module has been finished and the engine has come to a decision to execute a parallel QEP, the clone generation module can be started. The execution of this module begins with the runner agent creating clones (cloning is supported by several agents platforms such as IBM AGLETS). Hence, this module is executed by the runner agent at the origin host. One clone for each query fragment is created, where the clone is responsible for executing a given query fragment. Each clone is dispatched in a parallel fashion and asynchronously. Therefore, at a given moment, more than one clone for a mobile query can be running in different mobile databases (in different mobile hosts).

It is important to note that the Clone Generation module is activated only if the query engine has decided to execute a given QEP as a parallel QEP.

Fragment Execution Module

The Fragment Execution module is started when a runner agent (or its clones) migrates to the mobile units in which query fragments of a mobile query should be executed. On arriving at a mobile host, the runner agent (or a clone of it) submits to the local wrapper agent the query fragment (as an XQuery expression). The wrapper agent, in turn, translates the fragment into the native query language of the local database. Thereafter, the wrapper agent submits the query to the local database and returns the results back to the runner (or a clone). Obviously, local database systems can further optimize the processing of the query fragment.

Local statistical information can be gathered by the runner agent (or clone), while the Fragment Execution module is been executed. That is because information about operation, table and selection cardinalities, for example, can be inferred while the query fragment is being executed. In that case, when the runner (or clones of it) returns to the coordinator, it passes the collected statistical data to the wrapper agent.

As already mentioned, each operator node in a QEP can be annotated with triggers which execute actions when specified events happen. Such triggers might be executed during the execution of query fragments. During the execution of a fragment, some events may occur, such as, unavailable data sources or the expected number of tuples is different from the number of actually retrieved tuples. The occurrence of such events should trigger a reactive action. The runner agent is responsible for performing the action which can modify the local fragment or others fragments or even the whole query execution plan. Intuitively, it means the query execution can vary dynamically from full linear execution to full parallel execution.

Post-processing Module

This module is started when the runner agent and its clones (if there exist any) begin to arrive at the mobile host in which they were created (Coordinator). During the execution of this module, the clones send to the runner agent the results of the query fragments. The runner agent, in turn, starts the execution of (global) operations which could not be executed by any mobile database. In our approach, we use adaptive operators such as AJAX (see Section 4.3) to execute global operations.

4.2 *The Cost Model*

In this section, we present a cost model for the proposed query processing strategy. The cost model is based on statistical information about data stored in mobile databases and network transmission costs. *The cost is computed in bytes.*

The cost for processing a query (C_{MQ}) is estimated by the formula

$$C_{MQ} = \underbrace{\sum_{i=1}^{N}(Tx_{AGi})}_{A} + \underbrace{\sum_{i=1}^{N}(Tx_{AGi} + C_{Datai} + C_{Loci})}_{B} + \underbrace{\sum_{i=1}^{N}(C_{Gopi})}_{C}, \text{ where}$$

- **N** represents the number of mobile hosts in which fragments of the query have to be executed;
- $\mathbf{Tx}_{AGi}$ represents the transmission cost of a mobile agent i (in bytes);
- $\mathbf{C}_{Datai}$ represents the cost for transmitting the result of a query fragment executed at mobile unit i. This cost is calculated as follows:

  (number of tuples)(size of each tuple)*;

- $\mathbf{C}_{Loci}$ represents the cost (number of accessed pages) for processing a query fragment at mobile unit i, and it is calculated as follows:

  (number of pages)(size of a page in bytes)*;

- $\mathbf{C}_{Gopi}$ represents the cost for processing a global (high-level) operation.

In the formula, the component A indicates the cost for the first migration of one or more mobile agents. This cost is estimated when the query engine, during the execution of context recognition module, dispatches agents to one or more mobile units. In this case, that component captures the cost of transporting mobile agent codes. The component B is estimated during the execution of the fragment-execution module and represents the cost of agent migration with query result. Component C indicates the cost of processing a global operation at the mobile unit where the coordinator is running. Observe that the factors $\mathbf{C}_{Datai}$ and $\mathbf{C}_{Gopi}$ are estimated based on statistical information.

4.3 Adaptive Features

Traditional query processing techniques fail to support the access to databases in an MDBC, since data access (or arrival) in such an environment becomes unpredictable and mobiles hosts (in which mobile databases reside) may suffer from limited available main memory to process some query operators (e.g., join). The data access unpredictability arises due to the constants communication disruption, which mat occur frequently in wireless networks, and limited processing capability of some mobile hosts, which may cause a bursty arrival of tuples.

Adaptive query processing techniques [9, 10, 11, 15, 22] are quite adequate to be applied in dynamically changeable environments such as MDBCs. Those techniques carry out usefully work when an abnormal situation occurs. Our strategy integrates the use of adaptive operators and triggers techniques to cope with problems which may arise when executing queries in a dynamically configurable environment. In this section, an adaptive operator, denoted AJAX, is described. Furthermore, a strategy for processing queries with unavailable data sources is presented.

4.3.1 AJAX – An Adaptive Physical Operator

AJAX is the acronym of Adaptive Join Algorithm for eXtreme restrictions. We claim that AJAX is an adaptive query operator [21], since it is able to react to events which may considerably increase response time when executing queries over mobile databases. AJAX ensures the following properties:

(i) *Incremental production of results as soon as the data become available*. Conventional join operators (like nested-loop join and hash join) just produce results when at least one involved relation is completely evaluated. Joins executed over distributed data may take a long time until users begin to receive the result from the join operation. AJAX uses a pipeline technique in order to incrementally provide results (to users or to another query operator) as soon as they become available. By doing this, it is possible, for instance, to users (observing the progress in the query processing) to decide to stop the execution on-the-fly, since the results obtained are already enough to make a decision;

(ii) *Progress of the query processing even when the delivery of data is blocked*. The idea is to overlap delays by executing other tasks when tuple delivery (from tables involved in join operation) is blocked;

(iii) *Reaction to memory-overflow events*. When a memory overflow occurs AJAX triggers a memory-management policy which consists of flushing part of the data from main memory to disk;

(iv) *Prevention of unnecessary disk access*. Adaptive hash join operators (e.g. XJoin [22]) typically flushes data to disk when memory overflows occur. In order to avoid incomplete results and/or duplicates, data flushed to disk may be read (accesses) more than once. AJAX is designed to reduce such disk accesses by avoiding on-the-fly duplicate results (during probing time).

An adaptive query operator should be able to react to some events, which are related to increasing response time on mobile databases access. In order to explain the idea of adaptive operator, consider that a query Q consisting of a join operation has been submitted, and at undetermined running time one of the data sources (which is an operand of the join operation) has suddenly block its tuple delivery for some unknown reason. In other words, the join operation is blocked, and the query's response time is increased until the blocked data source resume to delivery tuples again to the join operation. This scenario is quite common in mobile databases and conventional operators are not prepared for such a situation.

AJAX keeps running even when both data sources (operands of the join operation) are blocked. The idea is to overlap delays by executing other tasks. That is achieved because AJAX is a symmetric operator and implements intra-operator pipeline as we show next.

AJAX's First Phase

The key goal of the first phase is to join the largest amount of memory-resident tuples, using available memory. This phase starts as soon as tuples begin to arrive and continues executing as long as tuples of at least one of the tables (involved in the join operation) are arriving. When all tuples of both tables have arrived, the first phase stops its execution (and the second phase is triggered). If the arrival of tuples from both tables is interrupted (blocked), the execution of the first phase is suspended (and the second phase is started), provided that there are no new tuples to be processed. When new tuples restart to arrive, the execution of the first phase is resumed. For that reason, the AJAX is a non-blocking join operator.

When the first phase starts its execution, two hash tables are created (one for each table of the join operation). Each hash table is composed by buckets. The hash tables are populated by applying a hash function ***h*** over the join attribute of tuples which arrive to the operator. AJAX's hash tables are double ended queues (deque) of buckets. In turn, each bucket is a linked list of tuples. There is no size limit for a bucket, which characterizes it as a dynamic data structure. The feature of having dynamic buckets is a way to implement a more fair memory allocation policy, since all available memory is used to allocate as many tuples as possible. Buckets with fixed size do not offer the best memory allocation. To illustrate this assertion, consider a fixed-size bucket which can hold at most 100 tuples, but at a given moment only 5 has already been allocated in it. Hence, when fixed-size buckets are used, one may have several empty (or almost empty) buckets holding memory area which could be used to allocate tuples belonging to other buckets. Besides that, using fixed-size buckets implies on memory overflow occurring more frequently, since the memory-overflow event occurs for buckets and there are several buckets for each hash table. Such a feature may represent a significant increase in query response time. In AJAX, a bucket that "deserves" to be larger (in size) will be in fact larger.

For each set of received tuples by AJAX, the memory size necessary to store the received tuples is calculated and compared with a minimal memory limit (MML), which is a memory size large enough to keep the algorithm running and to load disk pages to memory during the second phase. The minimal memory limit can be set automatically by AJAX depending on available memory. For example, at a given moment t there is enough room to allocate AJAX's code and four disk pages, but at moment k, with $t < k$, there is room to allocate AJAX and ten disk pages. Thus, at moment t, MML is set to m, while at moment k, AJAX can set MML to α, where $\alpha > m$. Such a feature makes AJAX adaptable to the amount of available memory. After checking if there is enough room in memory, the tuples are allocated in their corresponding hash tables by applying a hash function on the join attribute. The hash function result is the bucket address which will receive that particular tuple. Observe that the overflow granularity in AJAX has been moved from bucket level to system (memory) level. Observe that overflows occur less frequently at system level than at bucket level.

AJAX has been designed to do as many probing as possible in a specific time interval. To achieve this goal, AJAX implements inter and intra-operator pipelining. Some additional information is required in order to avoid duplicates and unnecessary probing. First, for each received tuple AJAX adds an index (a new column), which marks the tuple's relative position in the bucket. In other words, this index represents the insertion order of a tuple in the bucket. Second, AJAX also adds a column called *lastprobe remembrance*, which has the functionality of identifying the last tuple on the opposed bucket that has been probed with a given tuple. This information will be used to guarantee the *progressive probing* correctness and completeness.

Probing

Probing in AJAX is progressive; this progressiveness avoids duplication and unnecessary probing. The idea is to run the probing phase as a loop, which compares all pairs of buckets (from both relations) with the same hash address, comparing tuples that have arrived during the current iteration ***i*** and tuples from iteration ***i-1***, i.e., tuples that have arrived during the last iteration but have not been probed yet .

To illustrate the concept of progressive probing, consider the execution scenario depicted in Fig. 3. Observe that the join operation involves relations alpha and beta. For each hash table, there are two buckets (x and y). In an iteration i, the probing is done by comparing tuples from buckets with the same hash address; tuples belonging to the bucket with hash address "x" of alpha are compared with tuples of the bucket "x" of the beta's hash table, producing result-tuples incrementally. This procedure is executed for every pair of buckets in the hash tables until the last pair is reached. After that, a new iteration ***i+1*** for the probing phase starts from the beginning of the bucket list. For that reason, a deque data structure has been chosen; the algorithm's loop traverses all buckets of the bucket list in each iteration. Beginning from the list's head guarantees that tuples that have been received during a given iteration i may be probed in the next iteration ***i+1***, since some tuples may be inserted in bucket x from alpha while the probing is in progress in the bucket pair (from both hash tables) with hash address y, where $x > y$. Therefore, tuples which have arrived during the iteration ***i*** may or may not be probed during this iteration.

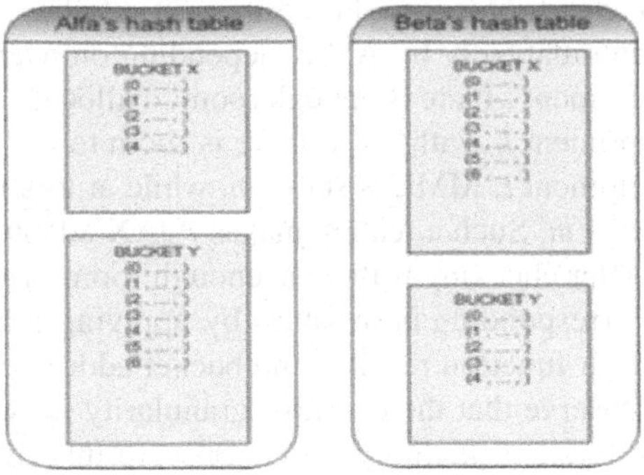
Alfa's hash table
BUCKET X
BUCKET Y
Beta's hash table
BUCKET X
BUCKET Y

Fig. 3 Running scenario of progressive probing

However, AJAX guarantees that in such a scenario, those tuples will be probed in next iteration ***i+1***. For that reason, the probing phase in AJAX is called progressive.

In order to avoid duplicates in the result of the join operation and unnecessary probing, AJAX implements the following strategy. Whenever a tuple is received by AJAX, it adds two columns to the tuple. One column, called position index, represents the receive order of the tuple in a given bucket. The other column, denoted *lastprobe remembrance* (LPR for short), has the following semantics: the LPR of a tuple ***t***, which belongs to a bucket with hash address x, stores the value of the position-index column of the last tuple of the bucket with hash address x of the other hash table which has already been compared (probed) with ***t***. Thus, for each tuple t. in the bucket with hash address x of Alpha's hash table, AJAX compares its LPR with the position index of all tuples in bucket x of the other hash table, say Beta. Two situations are possible:

- *(i)* The value of LPR of Alpha's tuple is less than the value of position index of Beta's tuple. In that case, the tuples have not been probed yet. For that reason, their join attributes are compared to check if the tuples should be added to the result. After that, The LPR of Alpha's tuple receives the position index value of the Beta's tuple. This guarantees that those tuples will not be probed any more.
- *(ii)* The value of LPR of Alpha's tuple is greater or equal to the value of position index of Beta's tuple, which means that the tuples have already been compared, and will not be probed again.

Fig. 4 illustrates how AJAX uses the LPR and position-index values. The last attribute of every tuple in Fig. 4 represents the LPR column. On the other hand, the first column represents the position-index value. Observe that the first tuple belonging to bucket 'x' of Alpha's hash table has position-index equal to 0 and LPR equal to 2, which means that it has already been probed with the tuple of position-index with values greater than -1 and less than 3 in Beta's bucket. Tuples with LPR equal to -1 are new tuples (recently allocated), which were not compared with any other tuple.

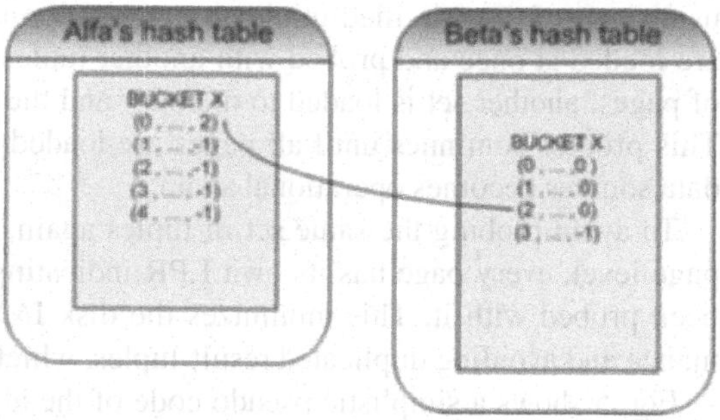
Alfa's hash table
BUCKET X
(0,…,2)
(1,…,-1)
(2,…,-1)
(3,…,-1)
(4,…,-1)
Beta's hash table
BUCKET X
(0,…,0)
(1,…,0)
(2,…,0)
(3,…,-1)

Fig. 4 Last Probe Remembrance

Reacting to memory overflow

AJAX reacts to the occurrence of memory overflow events. Such a property is possible, because AJAX checks the amount of available memory at tuple arrival time. Whenever new tuples arrive, AJAX calculates the required room to allocate those tuples. If there is enough area to allocate them, then the new tuples are inserted into their corresponding buckets. Otherwise, the algorithm makes room to allocate the incoming tuples. In order to free memory area, some buckets should be deallocated. Several deallocation policies can be implemented by AJAX. The default policy chooses the largest bucket pair (the pair with the largest quantity of tuples). After that, the algorithm flushes those buckets to disk and deallocate all tuples stored in those buckets. This is a recursive process, which is executed until there is enough memory area for allocating incoming tuples. When both data sources are blocked or EOF is received from them, the AJAX's second phase is triggered.

AJAX's Second Phase

Second phase uses buckets previously flushed to disk to produce result-tuples. The execution of the second phase is triggered to react to different events. First, when both data sources are blocked (due to external events such as low bandwidth, communication disruption or abrupt shutdown), AJAX executes starts the execution of the second phase. Second, when all the tuples of both tables are received, the second phase is triggered for the last time in order to ensure completeness of the join result produced by AJAX.

It is worthwhile to note that the execution of the first and second phases is alternated whenever data sources become blocked. In order to continue generating result-tuples, the second phase uses the buckets flushed to disk to feed the progressive probing loop.

As already mentioned, buckets are flushed to disk to avoid memory overflow. However, those disk-resident buckets could produce another overflow when they are loaded to be probed with other buckets in memory. This "second type of memory overflow" is prevented by using the minimal memory limit. Before flushing buckets to disk, AJAX executes an overflow preventing algorithm, whose key goal is to divide each single bucket in smaller partitions having the same size of a disk page (4 KB, for example). After that, pages can be flushed to disk. The loading is done using part of the specified minimal memory limit (MML) area; only a few pages are loaded at once and probed with another tuples on memory. After probing a set of pages, another set is loaded to memory and the previously ones are deallocated. This process continues until all pages are loaded and probed, or when one of the data sources becomes operational again.

To avoid probing the same set of tuples again, the LPR concept is used at disk page level, every page has its own LPR indicating the last page that have already been probed with it. This minimizes the disk I/O, increasing the AJAX's performance and avoiding duplicated result-tuples, which leads to incorrect results.

Fig. 5 shows a simplistic pseudo code of the algorithm, many details have been omitted in order to preserve a didactic approach. It is important to say that the process of receiving tuples from data sources and the overflow preventing process are executed in different threads, which is not represented in Fig. 5.

```
Procedure TupleArrival (tuple t, sources (A, B))
Begin
1. If t exceeds the minimal memory limit
  (a) Choose two buckets Ax and Bx.
  (b) Probe buckets Ax and Bx.
  (c) Flush buckets Ax and Bx.
  (d) Deallocate Ax and Bx.
2. Calculate the hash of t.
3. Insert t in the correct bucket.
End
Procedure ProgressiveProbing (buckets (Ai, Bi), pages (PAi, PBi))
Begin
1. If buckets Ai or Bi got a new tuple t
  (a) Probe t with all tuples with position index lower than its LPR.
  (b) Update the LPR of all probed tuples with the position index of t.
  (c) Update the result stream.
  1.1 If sources are blocked
    (a) Probe Ai with all disk pages with index less than its LPR.
    (b) Probe Bi with all disk pages with index less than its LPR.
    (c) Probe page PAi with all disk pages with index less than its LPR.
    (c) Probe page PBi with all disk pages with index less than its LPR.
    (d) Update the LPR of the involved tuples.
    (e) Update the result stream.
End
```

Fig. 5 AJAX Simplified

Cost Estimation

In order to estimate the cost of executing the AJAX operator, consider the scenario depicted in Fig. 6. **M** represents a mobile database community (MDBC). DB_1, DB_2 and DB_3 represent the mobile databases belonging to **M**. Those mobile databases reside in mobile hosts C_1, C_2 and C_3, respectively. The mobile hosts C_1, C_2 and C_3 are interconnected by means of an ad hoc network. Now suppose that a user submits a query Q in C_1. The query Q represents in fact a join operation (of the relational algebra) between two unsorted tables A and B, which are stored in DB_2 and DB_3, respectively. The tuples of relations A and B should be delivered to the mobile host C_1 where the join operation has to be executed. Thereafter, the result of Q should

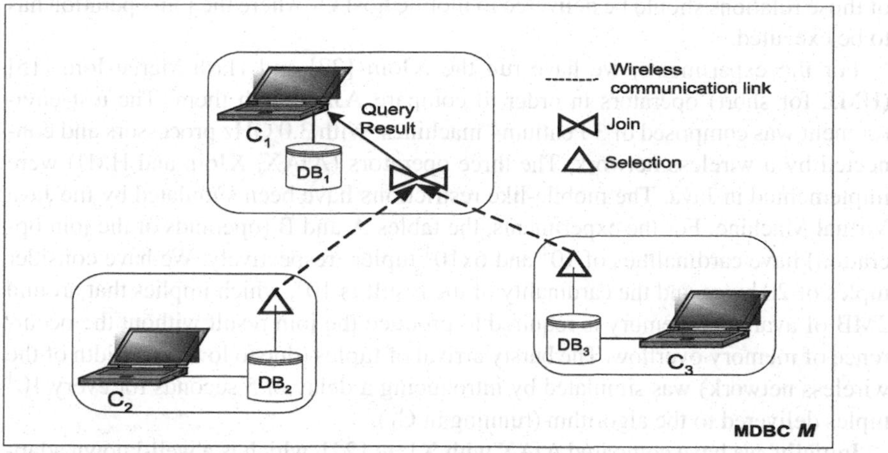
Wireless communication link
Join
Selection
Query
Result
C1
DB1
DB2
C2
DB3
C3
MDBC M

Fig. 6 Running Example

be delivered to the user, who has submitted Q. We are assuming that the database systems in C_1, C_2 and C_3 are interoperable.

For the sake of simplicity, let us consider that the tables A and B have the same cardinality c, the same tuple size and there is an even distribution of tuples between the two hash tables. The available main memory area in C_1 for executing the AJAX has the capacity of storing s tuples. Without lost of generality, the measure for estimating that cost is the number of tuple transfers to/from disk, instead of block (page) transfer. The cost for AJAX can be estimated, according the following conditions:

1. $s \geq 2c$. In this case, there is no overflow. Thus, no tuple is flushed to disk and the third phase is not triggered, which gives a cost of **2c** (for reading tables A and B);
2. $\frac{s}{2} < c \leq s$. In this case, there is an overflow of c tuples, since we have assumed an even distribution of tuples across the hash tables. The cost is given by $2c + (\frac{c}{2} + \frac{c}{2}) + (\frac{c}{2} + \frac{c}{2})$, i.e., $4c$. Note that, one term $(\frac{c}{2} + \frac{c}{2})$ represents the cost of flushing tuples to disk and the other represents the cost of reading the tuples from disk for executing the third phase.
3. $s < c$. In this situation, we have an overflow of $(2c - s)$ tuples, i.e., an overflow of $(c - \frac{s}{2})$ tuples for each table. Thus, the estimated cost is $2c + (2c - s) + (2c - s)$, giving the final cost of $2(3c - s)$

4.3.2 Experimental Results

In order to investigate performance of the AJAX operator, we have performed several experiments. In the experiments, a query Q is submitted to a mobile computer C_1, where Q represents a join operation between two unsorted tables A and B, each of which residing in a different mobile computer, C_2 and C_3, respectively. The tuples of those relations should be delivered to mobile host C_1 where the join operation has to be executed.

For the experiments, we have run the XJoin [22] and Hash-Merge-Join [15] (HMJ, for short) operators in order to compare AJAX with them. The test environment was composed of 3 Pentium4 machines, with 3.0 GHz processors and connected by a wireless network. The three operators (AJAX, XJoin and HMJ) were implemented in Java. The mobile-like restrictions have been simulated by the Java Virtual Machine. For the experiments, the tables A and B (operands of the join operation) have cardinalities of 10^4 and $6x10^4$ tuples, respectively. We have consider tuples of 20 bytes and the cardinality of the result is 10^5, which implies that around 2MB of available memory is required to produce the join result without the occurrence of memory overflow. The bursty arrival of tuples (due to low bandwidth of the wireless network) was simulated by introducing a delay of 5 seconds for every 10^4 tuples delivered to the algorithm (running in C_1).

Initially, we have compared AJAX with XJoin [22], which is a well-known adaptive join algorithm. Since disk access is the major bottleneck for join algorithms, it has been chosen as the quantity to be measured in the experiments. The results are illustrated in Fig. 7 and Fig. 8. Observe that both algorithms share a similar behavior until $6x10^4$ result-tuples are generated, which is the point when both algorithms have already flushed a great number of tuples to disk and more disk I/O is required to continue the operation. Looking at the AJAX series, one can see that the algorithm curve is quite linear even after $6x10^4$ result-tuples have been generated, which is not the case of XJoin (see Fig. 7). XJoin presents an exponential growth after $6x10^4$ result-tuples have been generated, which represents much more disk I/O than AJAX.

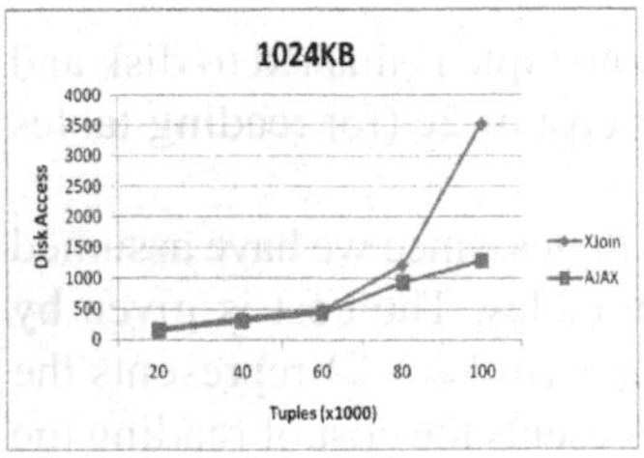
1024KB
Disk Access
4000
3500
3000
2500
2000
1500
1000
500
0
20
40
60
80
100
Tuples (x1000)
XJoin
AJAX

Fig. 7 1024KB of available memory

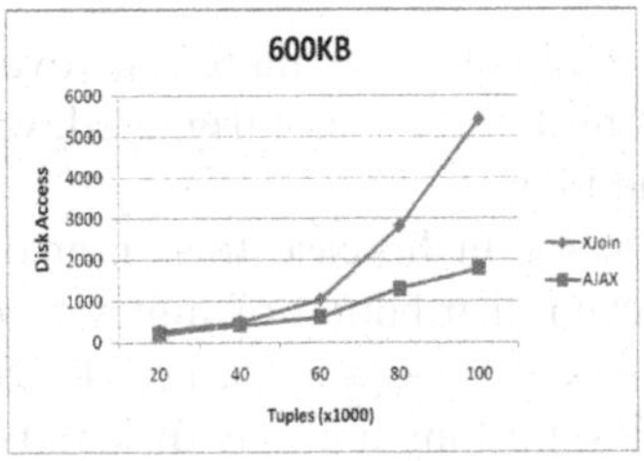
600KB
Disk Access
6000
5000
4000
3000
2000
1000
0
20
40
60
80
100
Tuples (x1000)
XJoin
AJAX

Fig. 8 600KB of available memory

As the available memory decreases, the performance similarity between XJoin and AJAX cease earlier, which can be observed in Fig. 8. In Fig. 8, one can also observe that to produce $4x10^4$ result-tuples XJoin increases its disk I/O exponentially, while AJAX increases its own disk I/O in a quasi-linear way (for the used cardinalities).

Besides the fact that disk I/O in XJoin grows faster as the available memory decreases, XJoin is machine-precision dependant. Since XJoin applies a timestamp to the recently arrived tuples, in order to avoid duplicates, it might still generate duplicates, because the timestamp can be duplicated if the machine in which the algorithm runs has not the required precision to count time. Considering that XJoin runs in a device capable to count time up to milliseconds, it cannot guarantee that two tuples that have arrived almost at the same time (in the same milliseconds tick) will receive different timestamps. For the sake of clarity, let tuple t_i be received at the real time of 12:33:0001 and tuple t_k is received at 12:33:0003, since the given device can only count up to milliseconds, the given tuples will receive the same timestamp of 12:33:000. Such machine-precision dependant behavior of XJoin is even worse when it is executing in mobile devices, since such devices may run processors with low time precision and no floating point support.

Another set of experiments have been performed, but this time, the Hash-Merge Join (HMJ) algorithm has been chosen to be compared with AJAX. First, we have simulated devices with 1024KB, 800KB and 600KB of free memory. For devices with 1024KB and 800KB, HMJ slightly produced less disk access than AJAX. However, such behavior changed when the available memory is shifted to 600KB: HMJ performance slightly degraded.

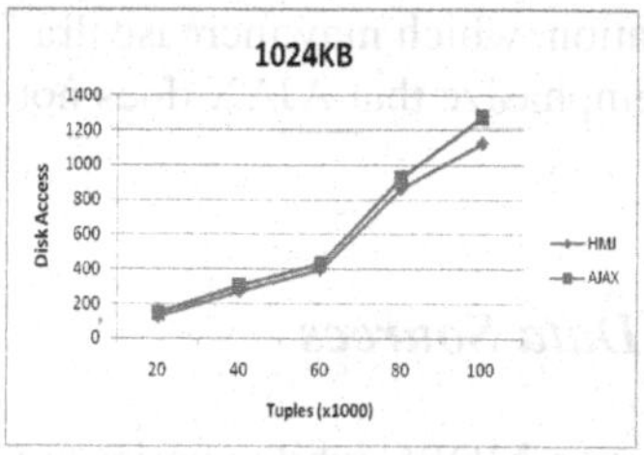
1024KB
Disk Access
1400
1200
1000
800
600
400
200
0
20
40
60
80
100
Tuples (x1000)
HMJ
AJAX

Fig. 9 AJAX x HMJ with 1024KB of available memory

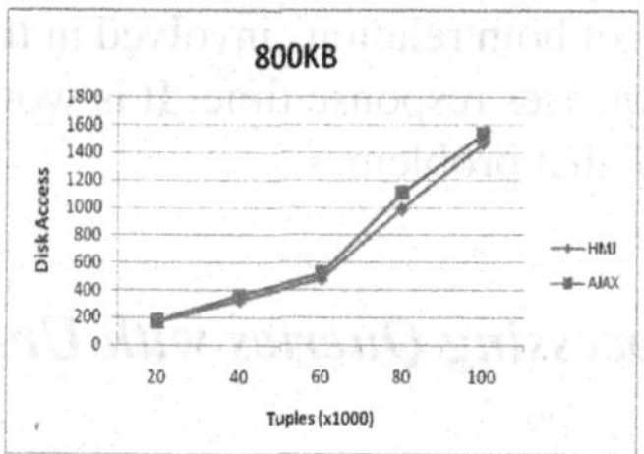
800KB
Disk Access
1800
1600
1400
1200
1000
800
600
400
200
0
20
40
60
80
100
Tuples (x1000)
HMJ
AJAX

Fig. 10 AJAX x HMJ with 800KB of available memory

From 800KB to 600KB, HMJ accessed the disk more often than AJAX at higher amounts of available memory. In order to analyze such behavior, the memory has been dropped to 512, 256 and 128KB of available memory, and the same behavior has been observed. Based on the data gathered from those tests and closer observation of the algorithms, one can conclude that such behavior is generated by the minimal memory limit imposed by the AJAX algorithm; that limit guarantees that no "second overflow" will occur when the algorithm starts loading disk buckets (pages) to the main memory. This "second overflow" is not handled by HMJ, which always use all memory available while running, which leads to another overflow while trying to load disk pages to compare with tuples still allocated in memory. From 600KB to 128KB, AJAX performance continued to increase (see Figures 11 and 12) compared to HMJ, even using a minimal memory limit as low as 16KB, which can allocate 2 disk pages (8KB) and handle the algorithm overhead (around 8KB).

It is important to note that HMJ requires the execution of an in-memory sort algorithm on its both phases. Of course, such sort procedure has a computational cost, which impacts in HMJ's overall performance. To estimate the sort cost in HMJ, let

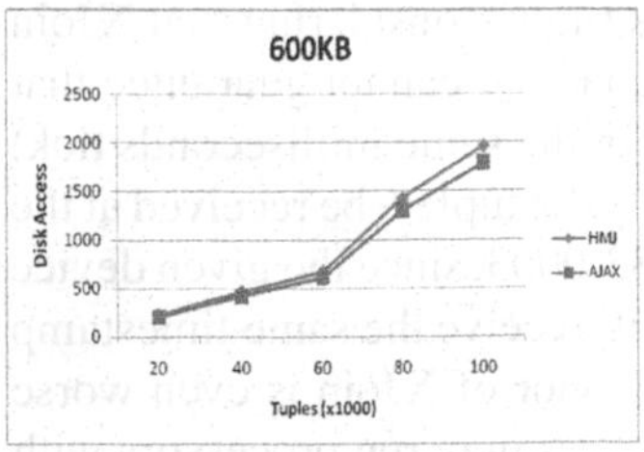
600KB
2500
2000
1500
1000
500
0
Disk Access
20
40
60
80
100
Tuples (x1000)
HMJ
AJAX

Fig. 11 HMJ performance degrades (with 600KB of available memory)

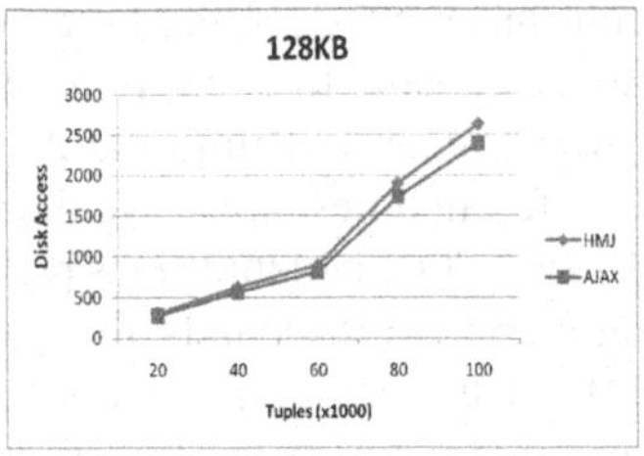
128KB
3000
2500
2000
1500
1000
500
0
Disk Access
20
40
60
80
100
Tuples (x1000)
HMJ
AJAX

Fig. 12 HMJ performance degrades (with 128KB of available memory)

k be the number of buckets of each hash table and n be the cardinality of each relation involved in the join operation. Considering that tuples of both relations are uniformly distributed among the buckets of both hash tables, the cost (complexity) for sorting the tuples in a given bucket is of $\frac{n}{k}\log\frac{n}{k}$, thus the cost for sorting all tuples during the execution of HMJ is $2n\log\frac{n}{k}$ .Observe, that cost depends on the cardinality of both relations involved in the join operation, which may increase dramatically queries response time. It is worthwhile to emphasize that AJAX does not suffer from that problem.

4.4 Processing Queries with Unavailable Data Sources

As already mentioned, disconnection of participants in an MDBC while queries are being processed may occur frequently. Therefore, during the execution of a global query, one or more data sources can be temporarily unavailable. In order to cope with unavailable data source problem, the proposed query processing mechanism uses the concept of E[C]-A rules.

The key idea of using E[C]-A rules is the following. When an "unavailable data source" event is detected, the query engine triggers an action on the QEP. There will be two possible actions to be triggered. First, if a linear QEP is being executed, the query engine reformulates the defined route. For example, if the unavailable data source represents the third mobile unit in the route list of the runner agent, the query engine reformulates the plan by sending a message to the runner agent informing that the third mobile unit is not available. This induces the runner agent to visit the other units first. After that, the runner should return to third unit and if it is available, the runner executes the query fragment. Observe that, in this case, most probably the reformulated plan is worse (i.e., it has a higher cost) than the initial plan.

Second, in the case of a parallel QEP, the query engine re-evaluates the impacted part of QEP. The original clone should be disposed. A new clone is created (and dispatched). The new clone is responsible for executing the query fragment again. If more than one data source is unavailable, the execution of the operations over the unavailable data sources should be changed to linear QEP. The other clones responsible for executing query fragments over available data sources continue their execution.

5 Processing Mobile Transactions

Users interact with database systems by means of application programs. A transaction is an abstraction that represents a sequence of database operations, resulting from the execution of an application program [1, 3, 17]. The component in the database system responsible for synchronizing the operations of concurrent transactions is the scheduler. The scheduler synchronizes operations belonging to different transactions by means of a concurrency control protocol.

Concurrency control protocols may present aggressive or conservative behavior [1, 17]. An aggressive concurrency control protocol tends to avoid delaying operations, and tries to schedule them immediately. However, such a strategy may lead to a situation in which there is no possibility of yielding a correct execution of all active transactions. In this case, operations of one or more transactions should be rejected. A conservative protocol, on the other hand, tends to delay operations in order to synchronize them correctly.

This section presents the Intelligent Transaction Scheduler (ITS) for synchronizing operations of concurrent mobile transactions submitted through AMDB. ITS is characterized by having a hybrid behavior (conservative or aggressive). The gist of this approach is to provide a scheduler that can automatically identify the modifications in the computational environment in which it is inserted by adapting its behavior without human interference. In order to prove the ITS efficiency, it was applied to control concurrency of transactions running in a Mobile Database Community.

5.1 ITS: A Self-adaptable Scheduler

The ITS is an intelligent scheduler composed of two modules (Fig. 13): the Analyzer, a component that makes decisions related to the most appropriate scheduler behavior; and the Scheduler, which executes the concurrency control protocols.

The Analyzer is an Expert System - ES based on fuzzy logic. Fig. 13 illustrates the basic components of the ES: *Knowledge Database*, which stores the specialized knowledge used in decision making; *Context*, which keeps the values of the variables used to solve the problem; *Inference Engine*, which infers knowledge by means of the rules stored in the Knowledge Database and the variable values stored in the *Context*; and *Interface*, which controls the dialog between the user (Scheduler) and the ES.

The problem that the Analyzer should resolve is to choose the most appropriate scheduler behavior, either aggressive or conservative, according to a number of computing environment characteristics. The most appropriate behavior should

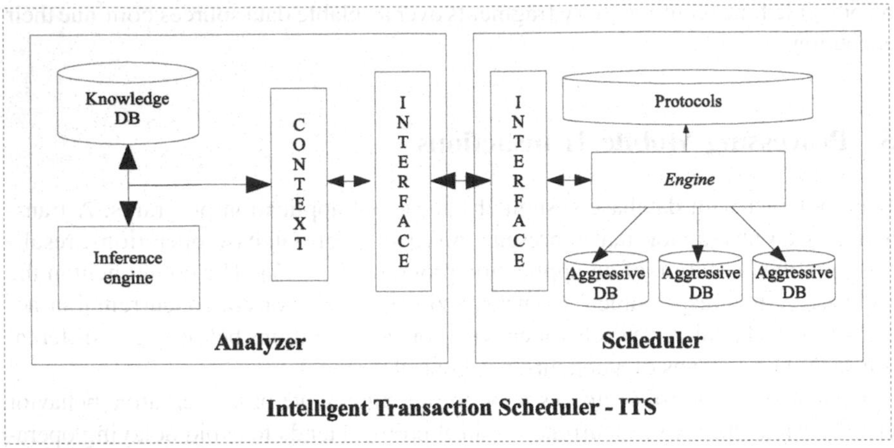
Knowledge DB
Inference engine
CONTEXT
INTERFACE
INTERFACE
Protocols
Engine
Aggressive DB
Aggressive DB
Aggressive DB
Analyzer
Scheduler
Intelligent Transaction Scheduler - ITS

Fig. 13 ITS Abstract Model

reduce the delays caused by the synchronization of the operations by the scheduler, while keeping the aborted transaction rate within the lowest range possible.

During the knowledge acquisition phase, four input variables were chosen: *aborted transaction rate* – *atr*, the percentage of the number of aborted transactions in relation to all transactions; *conflicting operation rate* – *cor*, the percentage of the conflicting operations in relation to the total number of operations; *lock waiting rate* – *lwr*, the percentage of the number of operations waiting for the lock to be released in relation to all operations of the conflicting transactions; and the *initial scheduler behavior* – *isb*. The output variable is called *scheduler behavior* – *sb*, which defines the most appropriate behavior for the adaptive scheduler.

The fuzzy sets for the input variables *atr*, *cor*, *lwr* were defined with a trapezoidal shape (Fig. 14). All these variables have the fuzzy values L (low), M (medium) and H (high), and their universe of discourse ranges from 0 to 100. The points x_1, x_2,

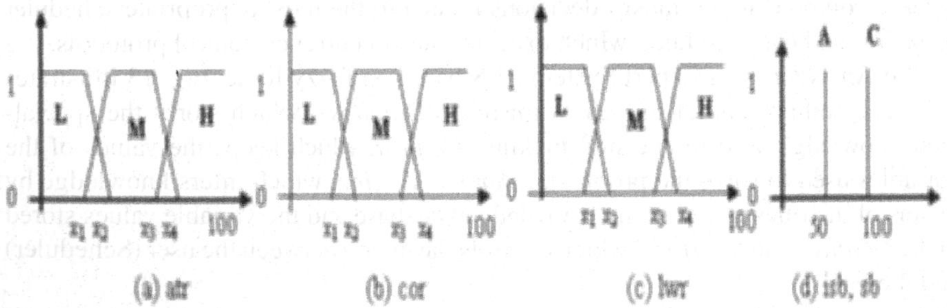
(a) atr
(b) cor
(c) lwr
(d) isb, sb

Fig. 14 Fuzzy Set for each linguistic variable

x_3 and x_4 define the values L, M and H. These points should be provided by the application/database administrator (a specialist).

The trapezoidal shape was chosen for these fuzzy sets because this shape has good response time (linear function) and its configuration is easy. Fig. 14d illustrates the fuzzy sets for *isb* and *sb* variables, where the singleton function has just two values: A (aggressive) or C (conservative).

The knowledge acquisition process for the definition of the rules was developed by means of many meetings between specialists. The observation technique was used to analyze fixed behavior schedulers. The knowledge database was structured into 8 rules as follows.

RULE 1: IF atr=L AND cor=L AND lwr = H THEN sb = A
RULE 2: IF atr=L AND cor=L AND lwr = M THEN sb = C
RULE 3: IF atr=L AND lwr = L AND isb = C THEN sb = C
RULE 4: IF atr=L AND isb = A THEN sb = A
RULE 5: IF atr=L AND cor = M AND isb = C THEN sb = C
RULE 6: IF atr=L AND cor = H AND isb = C THEN sb = C
RULE 7: IF atr=M THEN sb = C
RULE 8: IF atr=H THEN sb = C

The module of the ITS responsible for synchronizing operations is called *Scheduler*. The scheduler is composed of the following components: *Protocols*, which stores the code of concurrency control protocols; *Engine*, which executes the concurrency control protocols; *Aggressive DB* and *Conservative DB*, which store necessary information for scheduling the transactions when the scheduler has aggressive or conservative behavior, respectively; *Transition DB*, which stores necessary information for controlling concurrency of transactions during the transition period from one behavior to another; *Interface*, which implements the communication between the scheduler and the other components.

The ITS scheduler has three states: *Conservative State*, *Aggressive State* and *Transition State*. During the conservative and aggressive states all the transactions are scheduled using a concurrency control protocol with conservative or aggressive fixed behavior, respectively. The transition state, represents the transition from one behavior to another. In Section 5.2, we prove that ITS generates correct schedules during the transition phase.

5.2 *Ensuring Self-adaptability*

The key feature of the ITS stems from the ability it owns for adapting its behavior dynamically in response to changes in the computational environment. Consequently, ITS is able to switch the concurrency control protocol on-the-fly (i.e., without stopping its execution) and without human interference. Thus, the challenge ITS faces is to maintain the database in a consistent state during the transition phase, which corresponds to the time interval when there are transactions being synchronized by two different concurrency control protocols. The transition phase begins with the notification of behavior changing, which should be sent by the Analyzer to the Scheduler. This period ends when transactions, which were active (Definition 5.1) at the time of notification, become completed transactions (Definition 5.2).

Definition 5.1 (Active Transaction). *a transaction* T_i *is active if* $\forall o \in OP(T_i), o \neq a \wedge o \neq c$

Definition 5.2 (Completed Transaction). *a transaction* T_i *is completed if* $\exists o \in OP(T_i), o = a \vee o = c$

Before defining the notion of ITS schedules (schedules produced by the ITS), it is necessary to define a new operation over schedules, which is named *ordered join* (Definition 5.3). The ordered join of two schedules, $S \parallel S'$, will result in a new schedule S^J, which contains an interleaved sequence of operations belonging to schedules S and S' and the order of the operations belonging to S and S' is preserved in S^J.

Definition 5.3 (Ordered Join). *an ordered join of two schedules S and* S', $S \parallel S'$, *results in a schedule* S^J, *where:*

(*i*) $OP(S^J) = OP(S) \cup OP(S')$
(*ii*) $\forall p,q \in OP(S), p <_S q$ *then* $p <_{S^J} q$
(*iii*) $\forall p,q \in OP(S'), p <_{S'} q$ *then* $p <_{S^J} q$
(*iv*) $\forall p,q \in OP(S) \cap OP(S')$, *if* $p <_S q$ *then* $p <_{S'} q$ *and* $p <_{S^J} q$.

The ITS schedule (Definition 5.4) should be composed of the schedule produced by a conservative or aggressive protocol, plus the schedule in the transition phase.

Definition 5.4 (ITS schedule). *Let* S_0 *be a schedule on a set of transactions* T^0 *and* S_i *a schedule on a set of transactions* T^i, *where* $T^0 \cap T^i = \emptyset, 0 < i \leq n$. *A schedule* S^{ITS} *is defined as follows:* $S^{ITS} = S_0 \parallel_{i=1}^{n} (S_{transition_i} \parallel S_i)$, *where* S_0 *is the initial schedule, before any change of the scheduler behavior,* $S_i, 0 < i \leq n$, *is the schedule produced by ITS with a new behavior. The schedule* $S_{transition_i}$ *is the schedule in the transition phase and i is the number of scheduler behavior changes. If the scheduler does not change its behavior,* $S^{ITS} = S_0$.

The schedule produced during a transition phase involves the active transactions at the instant a behavior change notification arrives, called T_{old}, and the transactions created during the transition phase, called T_{new}. Thus, the set of transactions involved in the transition phase is $T_{transition} = T_{old} \cup T_{new}$. The schedule in the transition period, named $S_{transition}$, is the projection of S^{ITS} over the set $T_{transition}$. While old transactions were scheduled according to a particular protocol, the new transactions are scheduled by another protocol. Therefore, operations in $S_{transition}$ are scheduled by two different protocols coexisting during the transition phase - a protocol P1 used to schedule operations belonging to transactions in T_{old}, and the new protocol P2, which schedules operations of transactions in T_{new}. Recall that P1 and P2 are based on serializability and present different behaviors, either aggressive or conservative. In order to guarantee that $S_{transition}$ is correct, Corollary 5.1 (described next) should hold.

Corollary 5.1. *The conflicting operations belonging to T_{old} and T_{new} should be synchronized in the same (serialization) order by both protocols (aggressive and conservative).*

Lemma 5.1. *Let $S_{transition}$ be a schedule produced by the ITS during the transition phase. Then $S_{transition}$ is a correct schedule.*

Proof. During the transition phase, the ITS maintains two schedules - the schedule S_{old} produced by scheduler with the old protocol over T_{old}, and the schedule S_{new} produced by new protocol over T_{new}. The schedules S_{old} and S_{new} are correct since they are produced by concurrency control protocols based on serializability correctness criterion. According to Corollary 5.1, the ITS (during the transition phase) requires that the conflicting operations belonging to the sets T_{old} and T_{new} (of transactions) are serialized in the same order. Thus, $S_{transition}$ is correct. □

Theorem 5.1. *If S^{ITS} is a schedule produced by ITS, then S^{ITS} is correct.*

Proof. Consider S^{ITS} ,a schedule created by ITS, where S^{ITS} is defined as follows: $S^{ITS} = S_0 \|_{i=1}^{n} (S_{transition_i} \| S_i)$. S_0 and S_i are generated by protocols (with aggressive or conservative behavior) that use protocols based on already proven correctness criteria (e.g, serializability). The schedule $S_{transition}$ is correct (Lemma 5.1). Therefore, the serialization graphs for the schedules S_0, $S_{transition}$ and S_i do not have cycles. The schedules generated by ordered join operations in S^{ITS}, $S \| S_{transition}$ and $S_{transition} \| S_i$ are correct, that is, they do not introduce cycles in the SG(S^{ITS}), because (*i*) the order (of schedule) of the common operations on the schedules involved at an ordered join operation is preserved (item (*iv*) of (Definition 5.3), and (*ii*) S_0 and S_i are defined over two sets of transactions T^0 and T^i, where $T^0 \cap T^i = \oslash$, $0 < i \leq n$ (Definition 5.4). Since SG(S^{ITS}) does not have any cycle, S^{ITS} is correct. □

5.3 Experiments

In an MDBC, using a fixed concurrency protocol to generate correct schedules may not be the most efficient strategy. A scheduler with static behavior is not able to capture modifications in the context it is being executed. The arrival of a new mobile unit in an MDBC can drastically modify the scenario in which transactions are being executed.

There are two types of transactions in an MDBC environment: local transactions and mobile transactions. A local transaction is submitted directly to a mobile database on the same host. A mobile transaction, denoted M_i, consists of a set of subsequences $SUB_{i,1}$, $SUB_{i,2}$,. . ., $SUB_{i,n}$ of database operations, where each $SUB_{i,k}$ is executed at a mobile database MDB_k as an ordinary (local) transaction. In an MDBC, a local schedule models the execution of several interleaved operations belonging to local and mobile transactions in a given MU. A global schedule S represents the execution of all operations of the mobile and local transactions in the

MDBC [5]. Since each MU has autonomy on the execution of the local and mobile transactions, another component is necessary to guarantee the correct global schedule. In this context the ITS is responsible for scheduling the mobile transactions.

In order to guarantee correct global schedules using serializability, the necessary and sufficient conditions are: (*i*) every local schedule is serializable; (*ii*) there is a total order *O* over mobile transactions, such that at each MU the serialization order of mobile subtransactions is consistent with *O*. Since all existing database systems implement serializability, we assume that local schedules are always serializable. For the second condition, the ITS uses the implicit ticket method (ITM) [7].

In order to simulate the execution of ITS in an MDBC, we have assumed that an MDBC can be formed by *n* mobile units interconnected by a wireless network. Moreover, any MU can leave the network and new MU can join the MDBC. The implemented simulation environment has three basic components:

(*i*) MDBCSimulator. This component provides the user the support for defining the initial number of the mobile units, and the amount of available database items in each mobile unit, over which the database operations (i.e., read and write) are executed. By means of this component, the user can define when (in milliseconds) a mobile unit may leave or join the MDBC;

(*ii*) *TransactionSimulator*. This is a software component which creates mobile transactions. The operations of the mobile transactions are created randomly over database items defined by means of the MDBCSimulator. With the *TransactionSimulator*, the user may configure the number of transactions, the amount of database items accessed by the transactions and the percentage of read operations in each transaction;

(*iii*) ITS. This component represents the implementation of the proposed self-adaptable scheduler. For the simulation environment, the ITS implementation has two main modules: the *analyzer*, implemented with the JFuzzy API and the *scheduler*. For the scheduler module, the protocols TO (aggressive) and 2PL (conservative) were implemented.

To guarantee the necessary conditions for a correct global schedule in an MDBC, it is assumed that each mobile database implements a scheduler based on rigorous 2PL. This assumption ensures that each local schedule (produced by the local scheduler of a mobile database) is serializable and ITM can be used.

In order to use the ITS, the specialist should configure the fuzzy variable values used by the Analyzer. For all the tests presented herein, the following values were assigned: *atr* and *td* are considered low when between 0 and 20, medium when they vary from 40 to 60, and high when they vary from 80 to 100; and *ror* is considered low when its value is between 0 and 20, medium when it varies from 40 to 60, and high when it ranges from 80 to 100. The specialist ran many tests with aggressive or conservative schedulers for defining these values. This activity was carried out by using the implemented simulation environment.

In order to demonstrate that the ITS is able to identify changes and automatically decide on the most appropriate behavior, we simulated two scenarios in which initially all transactions were composed of read operations. After a while, a new

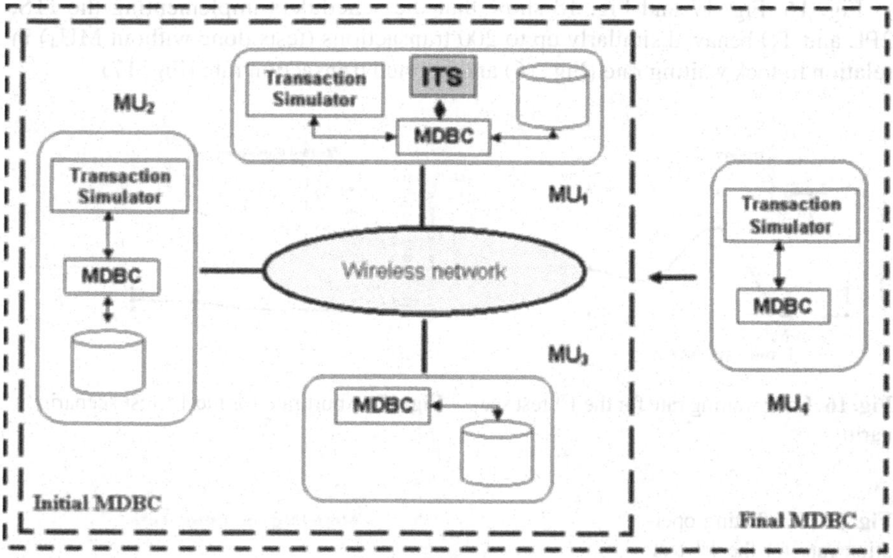
Transaction Simulator
ITS
MDBC
MU2
Transaction Simulator
MDBC
MU1
Transaction Simulator
Wireless network
MDBC
MU3
MDBC
MU4
Initial MDBC
Final MDBC

Fig. 15 Test Scenarios

mobile database joins to the MDBC, which submits mobile transactions composed of write operations.

First, it has been simulated an environment (Fig. 15), which is represents an MDBC, initially composed of 3 mobile units: MU_1, MU_2 and MU_3. After five seconds a new mobile unit, MU_4 joins the MDBC. Each mobile unit has an identifier. The MU_1 and MU_2 have a transaction simulator and a database composed of 20 available data items for the MDBC. The MU_3 has only one database. The MU_3 does not create mobile transactions. The mobile unit MU_4 has the transaction simulator, but does not have a database for the MDBC. The transactions generated by MU_1 and MU_2 are composed of read operations only and are executed over the global scheme of the MDBC. The mobile transactions generated by MU_4 are composed of 80% read operations (and 20% write operations). The write operations are executed over just one database item. The initial scheduler behavior is conservative. Five sets of tests were run in which the ITS synchronized operations of 100, 200, 300, 400 and 500 mobile transactions, respectively. In the tests with sets of 100 and 200 mobile transactions, the results obtained (Fig. 16, Fig. 17, Fig. 18) were derived from the transactions generated by MU_1 and MU_2 (50 and 100 transactions respectively). In the tests with 300, 400 and 500 transactions, the transaction simulators of the MU_1 and MU_2 generated 100 mobile transactions each, and the MU_4 generated 100, 200 and 300 mobile transactions, respectively. The tests were done in the following manner: for each set of tests, three different schedulers were used; the ITS, strict 2PL and TO.

Fig. 16, Fig. 17 and Fig. 18 show that the schedulers implementing the ITS, 2PL and TO behaved similarly up to 200 transactions (tests done without MU_4) in relation to lock waiting rate (Fig. 16) and aborted transaction rate (Fig. 17).

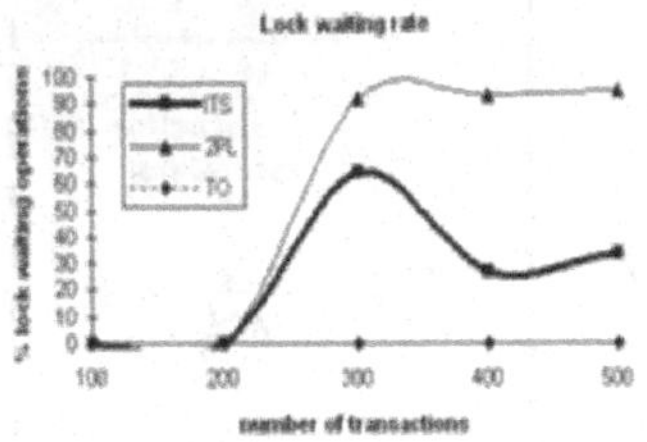
Lock waiting rate
ITS
2PL
TO
number of transactions

Fig. 16 Lock waiting rate for the 1st test scenario

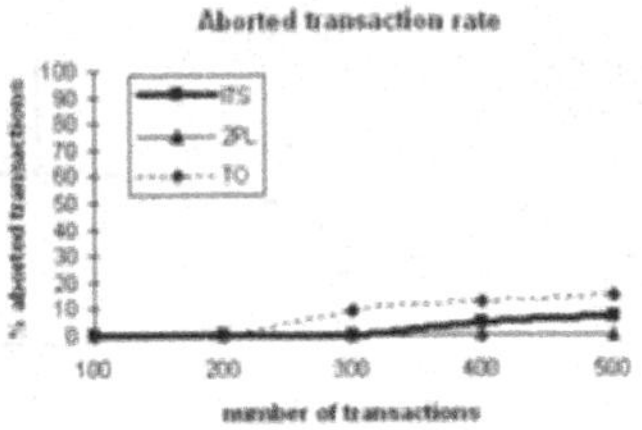
Aborted transaction rate
ITS
2PL
TO
% aborted transactions
number of transactions

Fig. 17 Abort rate for the 1st test scenario

Fig. 18 Conflicting operation rate for the 1st test scenario

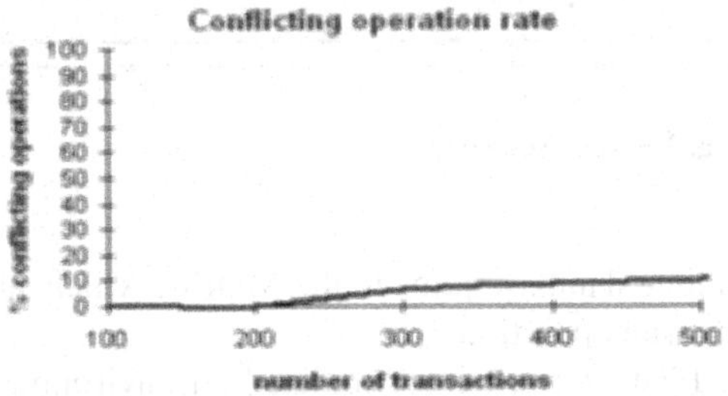
Conflicting operation rate
% conflicting operations
number of transactions

Under these conditions, as all transactions are composed of only read operations, there will be no conflicting operations, consequently there will be neither operations in queue, nor aborted transactions. After the new mobile unit, MU_4, entered the MDBC, the ITS demonstrated better performance than the 2PL in relation to the lock waiting rate (Fig. 16), maintaining the aborted transaction rate low (Fig. 17). In this scenario, the ITS initially began with conservative behavior (2PL) and changed to aggressive behavior (TO). The protocol that presented the worst performance in relation to lock waiting rate was the 2PL, because the mobile transactions generated by MU_4 have one write operation executed over only one data item (a type of hot spot), and there is a high number of operations waiting for the lock in this data item to be released. Note that in these tests the conflicting operation rate (Fig. 18) was low, lower than 20%, and the aborted transaction rate was between 0 and 20% (Fig. 17), during the entire test period. The ITS' Analyzer used Rule 1 to define that the aggressive behavior was the most appropriate. Looking more closely at Figures 16 and 17, one can observe that the ITS is able to find a good compromise between the abort rate and lock waiting rate compared to schedulers implementing a fixed concurrency control. Therefore, such results prove our claim that an adaptive scheduler is quite appropriate for controlling concurrency in dynamically configurable environments.

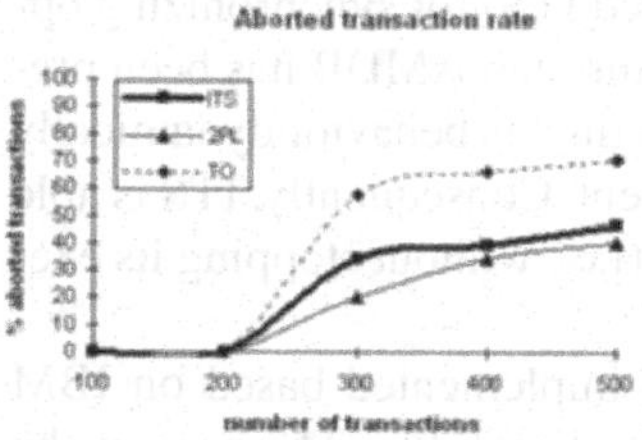
Aborted transaction rate
ITS
2PL
TO
% aborted transactions
100
90
80
70
60
50
40
30
20
10
0
100
200
300
400
500
number of transactions

Fig. 19 Abort rate for the 2nd test scenario

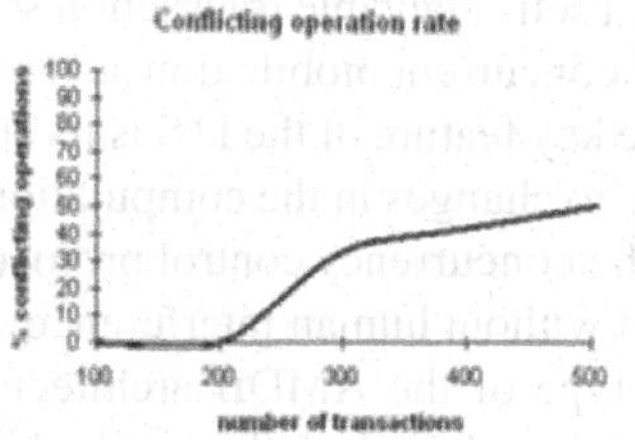
Conflicting operation rate
% conflicting operations
100
90
80
70
60
50
40
30
20
10
0
100
200
300
400
500
number of transactions

Fig. 20 Conflicting operation rate for the 2nd test scenario

After simulating ITS execution in the aforementioned environment, we have slightly modified that environment by introducing the following feature: transactions submitted by MU_4 are now composed of 20% read operations and 80% write operations. Those operations are executed over a set of 20 data items. In order to evaluate the ITS performance, the same five sets of tests were executed. Moreover, the ITS' initial behavior was set to aggressive. Fig. 18 shows that the aborted transaction rate of the three schedulers (ITS, 2PL and TO) were the same (tta = 0) up to 200 transactions, since all operations are read operations. After the MU_4 entered, the TO scheduler increased the aborted transaction rate, because the mobile transactions generated by MU_4 have many write operations, generating many conflicting operations (Fig. 20) and aborted transactions (Fig. 19). In these tests the ITS began with aggressive behavior and changed to conservative behavior. The ITS' Analyzer used Rule 7 to recognize that the conservative behavior was the most appropriate.

6 Conclusions

In this chapter, we define the concept of dynamically configurable database communities (MDBCs) in MANET environments. In order to provide a platform to enable MDBCs to be formed opportunistically over mobile database a middleware is described. The proposed middleware, denoted AMDB (Accessing Mobile Databases), supports physical mobility of hosts (in which reside databases systems) and logical mobility of database queries (or transactions) across mobile hosts. Moreover, the proposed architecture has important additional properties for coping with database mobility, such as: (*i*) it does not require changes to the core of the underlying database systems, and; (*ii*) it supports opportunistic creation of database communities over a varying number of mobile and autonomous database hosts.

A query engine for processing queries over mobile databases in an MDBC through the AMDB architecture has been described and discussed. An MDBC is an unpredictable and dynamically changeable environment. For that reason, the proposed query processing strategy is adaptive to the query execution context. An adaptive operator, denoted AJAX, and a strategy for processing queries with unavailable data sources are presented as well.

Finally, a self-adaptable transaction scheduler (called ITS) for synchronizing operations of concurrent mobile transactions submitted through AMDB has been presented. The key feature of the ITS is its ability for adapting its behavior dynamically in response to changes in the computational environment. Consequently, ITS is able to switch the concurrency control protocol on-the-fly (i.e., without stopping its execution) and without human interference.

A prototype of the AMDB architecture has been implemented based on IBM AGLETS, a Java-based platform which supports logical mobility. The core of the AMDB architecture currently consists of 8,000 (approximately) lines of Java code.

We conclude this chapter by emphasizing that sharing information among multiple heterogeneous, autonomous, distributed and mobile data sources stored in nodes of a MANET has emerged as a strategic requirement for several applications. Providing interoperability of databases in dynamically configurable environment represents a new challenge to the database community.

References

1. Bernstein, P., Hadzilacos, V., Goodman, N.: Concurrency Control and Recovery in Database Systems. Addison-Wesley, Reading (1987)
2. Bouganim, L., Fabret, F., Mohan, C., Valduriez, P.: A Dynamic Query Processing Architecture for Data Integration Systems. IEEE Data Engineering Bulletin 23(2), 42–48 (2000)
3. Brayner, A., Alencar, F.S.: A Semantic-Serializability based Fully-Distributed Concurrency Control Mechanism for Mobile Multi-database Systems. In: Proceedings of the 16th International Workshop on Database and Expert Systems Applications, Copenhagen, Denmark, pp. 1085–1089. IEEE Press, Los Alamitos (2005)
4. Brayner, A., Aguiar Moraes Filho, J.: Sharing Mobile Databases in Dynamically Configuration Environments. In: Eder, J., Missikoff, M. (eds.) CAiSE 2003. LNCS, vol. 2681, pp. 724–737. Springer, Heidelberg (2003)
5. Brayner, A., Aguiar Moraes Filho, J.: On Mobile Transaction Processing in Dynamically Configurable Mobile Database Communities. Journal of Parallel, Emergent and Distributed Systems 21, 199–213 (2006)
6. Dunham, M.H., Kumar, V.: Impact of Mobility on Transaction Management. In: Proceedings of the 1st International Workshop on Data Engineering for Wireless and Mobile Access, pp. 14–21. ACM Press, Washington (1999)
7. Georgakopoulos, D., Rusinkiewicz, M., Sheth, A.: Using Tickets to Enforce the serializability of Multidatabase Transactions. IEEE Transaction on Knowledge and Data Engineering 6(1), 1–15 (1993)
8. Grant, J., Litwin, W., Roussopoulos, N., Sellis, T.: Query Languages for Relational Multidatabases. VLDB Journal 2(2), 153–171 (1993)
9. Haas, P., Hellerstein, J.: Ripple Joins for Online Aggregation. In: Delis, A., Faloutsos, C., Ghandeharizadeh, S. (eds.) Proceedings of the ACM SIGMOD International Conference on Management of Data, Philadelphia, USA, pp. 287–298. ACM Press, New York (1999)
10. Ives, Z., Florescu, D., Friedman, M., Levy, A., Weld, D.S.: An Adaptive Query Execution System for Data Integration. In: Delis, A., Faloutsos, C., Ghandeharizadeh, S. (eds.) Proceedings of the ACM SIGMOD International Conference on Management of Data, Philadelphia, USA, pp. 299–310. ACM Press, New York (1999)

11. Kemper, A., Wiesner, C.: HyperQueries: Dynamic Distributed Query Processing on the Internet. In: Apers, P.M.G., Atzeni, P., Ceri, S., Paraboschi, S., Ramamohanarao, K., Snodgrass, R.T. (eds.) Proceedings of the 27th International Conference on Very Large Databases, Roma, Italy, pp. 551–560. Morgan Kaufmann, San Francisco (2001)
12. Lange, D.B., Oshima, M.: Programming and Deploying Java Mobile Agents with Aglets. Addison-Wesley, Massachusets (1998)
13. Macker, J.P., Corson, M.S.: Mobile Ad Hoc Networks and the IETF. Internet Engineering Task Force, MANET Working Group, http://www.ietf.org/html.charter/manet-charter.html
14. Manolescu, I., Florescu, D., Kossman, D.: Answering XML Queries over Heterogeneous Data Sources. In: Proceedings of the 27th International Conference on Very Large Databases, Roma, Italy, pp. 241–250. Morgan Kaufmann, San Francisco (2001)
15. Mokbel, M.F., Lu, M., Aref, W.G.: Hash-merge Join: A Non-blocking Join algorithm for Producing Fast and Early Join Results. In: Proceedings of the 20th International Conference on Data Engineering, Boston, USA, pp. 251–263. IEEE Press, Los Alamitos (2004)
16. Murphy, A.L., Pico, G.P., Roman, G.-C.L.: A Middleware for Physical and Logical Mobility. In: Proceedings of the 21st International Conference on Distributed Computing Systems, Phoenix, Arizona, USA, pp. 524–533. IEEE Press, Los Alamitos (2001)
17. Özsu, M.T., Valduriez, P.: Principles of Distributed Database Systems, 2nd edn. Prentice Hall, Englewood Cliffs (1999)
18. Patel, J.M.: Query Processing in Mobile Environments. In: NFS Wokshop on Context Aware Mobile Data-base Management (CAMM), Providence, Rhode Island, USA (2002), http://www.sice.umkc.edu/nsfmobile/wshop.html/JigneshMPatel.pdf (Cited January 24-25, 2002)
19. Roman, G.-C., Pico, G.P., Murphy, A.L.: Software Engineering for Mobility: A Roadmap. In: Finkelstein, A.C.W. (ed.) Future of Software Engineering. ACM Press, New York (2000)
20. Singhal, M.: Techniques for Building Large Relational Databases on Mobile Computing Systems. In: NFS Wokshop on Context Aware Mobile Database Management (CAMM), Providence, Rhode Island, USA (2002) (cited January 24-25, 2002)
21. Raman, V., Hellerstein, J.M.: Partial Results for Online Query Processing. In: Franklin, M.J., Moon, B., Ailamaki, A. (eds.) Proceedings of the 2002 ACM SIGMOD International Conference on Management of Data, Wisconsin, USA, pp. 275–286. ACM Press, New York (2002)
22. Urhan, T., Franklin, M.: XJoin: A Reactively Scheduled-Pipelined Join Operator. IEEE Data Engineering Bulletin 23(2), 27–33 (2000)
23. XML Schema, http://www.w3.org/XML/Schema
24. XQuery 1.0 Formal Semantics, http://www.w3.org/TR/query-semantics/

Database Architecture of Diagnostic System for Large Power Transformers

Liliana Byczkowska-Lipińska and Agnieszka Wosiak

Abstract. Supporting service staff is the most important task of expert system for technical state of electric device. Permanent data acquisition enables displaying current parameters and forecasts for the future. It is also recommendable for the system to provide remedy actions in case of a particular fault.

According to above - mentioned requirements, computer system for monitoring power transformers was designed and implemented. The project of a database is one of the most important elements, that must have been taken into consideration during the process of system implementation.

1 Introduction

Nowadays the information systems based on database technologies are essential to every field of life. One of them concerns diagnosing technical state of different kinds of devices and technical processes. The diagnostic systems carry out analysis of information related to the state and operation of particular components and subassemblies of the diagnosed devices to detect the initial stage of the failure or wear [19]. The realization of the diagnostic system is based on data received from measurements. To get the most precise analysis the expert systems are often implemented.

The main task of a expert diagnostic system of electric devices is to support technical stuff and to collect on - line and off - line information on running and responding of a diagnosed object. It consists of [22]:

Liliana Byczkowska-Lipinska
Technical University of Lodz, Institute of Computer Science, ul. Wolczanska 215, 90-924 Lodz, Poland
e-mail: lilip@ics.p.lodz.pl
Agnieszka Wosiak
Technical University of Lodz, Institute of Computer Science, ul. Wolczanska 215, 90-924 Lodz, Poland
e-mail: agnieszka@ics.p.lodz.pl

D. Zakrzewska et al. (Eds.): Meth. and Support. Tech. for Data Analys., SCI 225, pp. 219–238.
springerlink.com

- continuous on - line acquiring of automatically loaded data ,
- irregular off - line data acquiring (entered by a keyboard),
- collecting current information (on - line and off - line) and forecast results,
- an assessment of device conditions on the basis of decision trees and implementation of decision rules gathered by an expert system,
- presenting information on a current state of the device,
- recommendation of actions possible to be taken in a particular situation,
- creating reports in tables and charts,
- short term and medium term forecasts, which suggest the application of the maintenance procedures,
- long term forecasts providing assumptions future performance of the device.

As it results from the above - mentioned tasks, the correct design of the acquisition module and data store is one of the main element of the diagnostic system planning.

The mentioned problems are widely discussed basing on the system which cooperates with database. The system uses measurement data coming from large power transformers. It was designed for the company which concerned with electric energy supply.

2 Relevant Research

The feasibility of diagnostic expert system strongly depends upon the mechanisms of knowledge acquisition and maintenance. In the last years, a number of research projects have been devoted to power transformer on - line monitoring [11, 12, 13, 14, 15, 16, 17]. Although many steps ahead have been made, the problem in general is very hard. This is confirmed by the fact that a number of methods have been described in literature, but very few real applications were implemented.

One of the known examples is the Monitoring System MS 2000 [11]. It uses a process control system for the execution of the multiple tasks of the software, such as event controlled measured data acquisition and processing, data storage, visualization and communication. The measuring data can be recorded in the millisecond grid by means of the operating system. Data storage in the MS 2000 system is carried out in three stages. Initially, the data are held in the RAM memory of the industrial PC. It is to prevent a frequent access to the hard drive. After a time interval, the mean values for each channel are field in a historical database of the system. The data are field in accordance with their physical properties. In a third step, the user can choose an excerpt of the database which after being downloaded onto the office PC can be edited off - line with an operator - specific evaluation software or with MS Excel.

The TM100 transformer monitoring system consists of data acquisition unit and personal computer [12]. Every minute a sample of all measuring quantities is taken. After that, the physical quantities are calculated and all the signals can be visualized. Every hour the 60 data sets of all measuring quantities are compressed to only one data set. In many cases the maximum value of the 60 minute samples is taken.

There is also an archive containing minute values of all signals' values recorded in 30 days. After 30 days, the oldest data set is deleted. Using a database for archiving the samples or compressed data sets, increases the availability and avoids unintended manipulation of the data. Both archives can be exported into text files and processed by all commercially available software.

The demand for diagnostic systems is still very high. They can improve efficiency of devices and reduce costs. That is why new systems using new technologies and solutions are designed and implemented.

3 The Architecture of the Diagnosing System

The diagnostic expert system has been implemented on ANER3D 160000/220PN 220/110kV and RTdxP 125000/200, 220/110kV power transformers placed in a electrical substation in Piotrkow. The system has an open architecture. It enables adding almost unlimited amount of sensing elements used for different kinds of measurements of physical values and related measurement subsystems as well as state logging (from the digital inputs) of any number of auxiliary devices. It also enables the maintenance of greater number of large power transformers placed in one substation. Moreover the system discussed in this paper is to work with other systems installed on the transformers. The schema of the system's elements location together with other systems already working on a electrical substation is shown on a fig. 1. It was assumed the discussed system ought to work with a Substation Control System (SCS).

The following symbol notation is being used on the fig. 1:

- MB - measurement box (there is one box for each of the transformer),
- SC - stationery computer,
- DES - data exchange stationery computer,
- AC - administration computer,
- SCS - Substation Control System,
- M - modem,
- R - router.

According to the schema shown on figure 1, the system gets measurement data from the SCS system. Additional data, necessary for the diagnostic process, not gathered in the SCS system, are taken by the controller which communicates with LAN. All the information from the power transformer are processed and then sent from the controller placed in the transformer's vat to the LAN using fiber optic cable.

The data exchange stationery computer ensures data access safety. Moreover there was no need to make any changes in the Substation Control System, which has been working for several years. The SCS system and the diagnostic system cooperate by data exchange. The diagnostic system gets data from the SCS system and enables its own measurement data to the SCS system. The Elkomtech software read off measurement data sent by SCS system. Then data are saved on a hard disk in

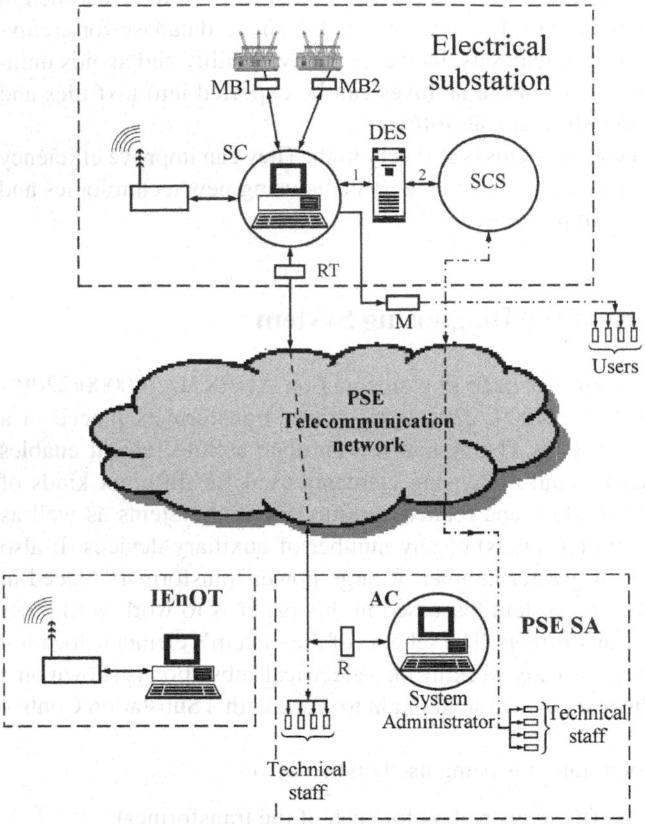
Electrical
substation
MB1
MB2
DES
SC
1
2
SCS
RT
M
Users
PSE
Telecommunication
network
IEnOT
AC
PSE SA
R
System
Administrator
Technical
staff
Technical
staff

Fig. 1 The schema of the system's elements location together with other systems working on a electrical substation

text files and are enabled to the diagnostic system. The diagnostic system's data are also saved in text files, which may be read by SCS system.

According to the requirements, the monitoring system mostly uses a meter - circuit Windex, which has already been implemented on a tasting station. The Windex system generates about five hundred signals for each of the transformers. It is necessary for the monitoring system to select generated signals and install the system on an intermediary computer. Additionally, it is necessary to take into consideration the fact that transformers have slightly different parameters and different sensing elements. That is why the signal sources must be treated individually while programming a system. As a result 101 binary signals and 25 meter signals have been chosen for each of the power transformer.

A block diagram for on - line meter circuits is shown on figure 2. Data coming from meter circuits are sent to a system computer by an optical fibre transmission

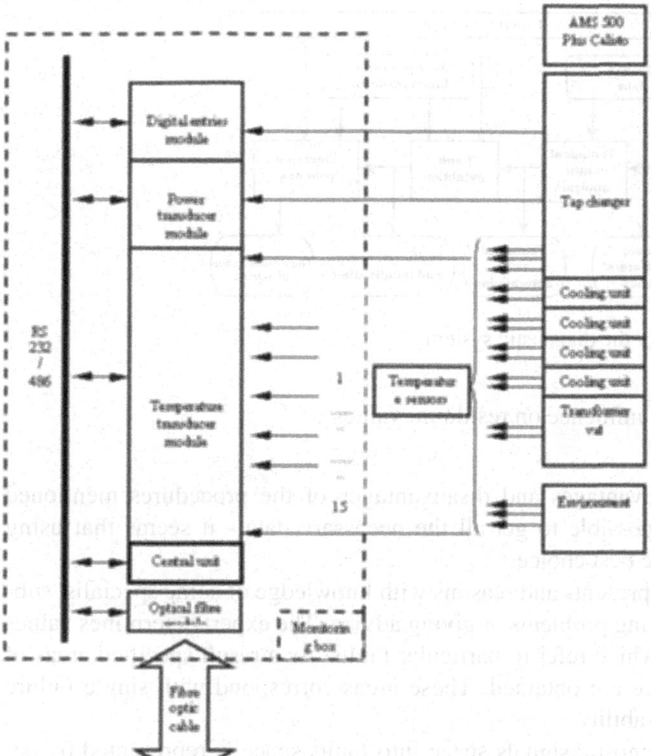
AMS 500
Plus Calisto
Digital entries module
Power transducer module
Tap changer
Cooling unit
Cooling unit
Cooling unit
Cooling unit
Transformer vat
Environment
RS 232 / 486
Temperature transducer module
1
15
Temperatur e sensors
Central unit
Optical fibre module
Monitorin g box
Fibre optic cable

Fig. 2 Transformer meter circuits

line and an optical fibre module. The system software uses data resources by entering text files which are periodically created on an intermediary computer.

The logical structure of the system is shown at figure 3. The logical structure of the described system is based on measurement data coming from meter circuits. Meter signals are subject to analysis and diagnosis while they are being loaded. The expert knowledge gathered in the knowledge base is the most important for this analysis. The off - line data are subject to analysis and diagnosis as well. After initial estimation the signals' values are displayed on a screen and the precise analysis takes place. It results in the detection of the state of the whole object or its parts. The knowledge base lets find out the fault location and the fault type. Then the recommendation of actions possible to be taken takes place.

Meter signals' analysis is determined by a specified model of diagnosis, failure localization and the state of the object detection. The main task of this model is to map a space of signal values into discrete space of faults or object states. The choice of the model is determined by [6]:

- learning,
- knowledge of redundancy of the hardware structure,

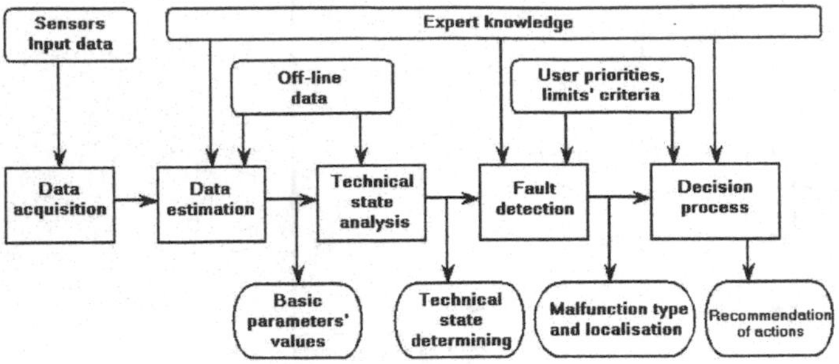
Sensors
Input data
Expert knowledge
Off-line
data
User priorities,
limits' criteria
Data
acquisition
Data
estimation
Technical
state
analysis
Fault
detection
Decision
process
Basic
parameters'
values
Technical
state
determining
Malfunction type
and localisation
Recommendation
of actions

Fig. 3 Logical structure of the diagnostic system

- modelling of faults' influence on residuum values
- expert knowledge.

Considering all the advantages and disadvantages of the procedures mentioned above - it is often impossible to get all the necessary data - it seems that using expert knowledge is the best choice.

An expert system represents and reasons with knowledge of some specialist subject with a view to solving problems or giving advice. The expert determines values of diagnostic signals, which refer to particular faults. As a result specified areas of diagnostic signals space are obtained. These areas correspond with single failure states and determine usability.

Mapping of the diagnostic signals space into faults space is represented by decision trees (or "if - then" principals) where each of the nodes corresponds with a number of brunches depending on values of a particular signal. The relationship between faults and values of diagnostic signals can be described in following statements [3]:

$$If\ (s_1 = \upsilon_{k1})\ and\ \ldots\ and\ (s_j = \upsilon_{kj})\ and\ (s_J = \upsilon_{kJ})\ then\ fault\ \varphi_k$$

where:

$$F = \{\varphi_k k = 1,2,\ldots,K\}\ - set\ of\ faults$$
$$S = \{s_j; j = 1,2,\ldots,J\}\ - set\ of\ diagnostic\ signals$$

Decision trees can represent any function mapping all values in attributes domain into category set - which means it can be any hypothesis. Decision trees representation is successful and enables highly efficient implementation of the process of the example classification (process of making use of hypothesis obtained by learning).

The efficiency of the example classification is of great importance to the system's speed and the comfort of the final user's work. Besides the effective usage of the memory is very important for the diagnostic systems, especially in these substations where there are limited costs for the computer devices.

Rules representation is also a very common method of inductive method. In condition part there are conditions to be or not to be fulfilled by examples and in decision part there is a category of examples that fulfils these conditions according to the hypothesis represented by the rule [30].

In knowledge systems based on a rule - based technique, the domain expertise to solve a problem is encoded in the form of "if - then" rule that are triggered by facts that match the if - part of the rule. Rules can be used to express heuristic knowledge, such as causal relationships, and to support deduction. When the antecedent of a rule is satisfied, the inference engine can infer the conclusion or execute the action. The implementation of the knowledge base and the expert system takes place by using MS Visual Studio .Net software tools.

An architecture of rule - based system include the following four basic elements [8]:

- Knowledge acquisition system. Knowledge acquisition is the accumulation, transfer, and transformation of problem - solving expertise from experts or documented knowledge sources to a computer program for constructing or expanding the knowledge base.
- Knowledge base. The knowledge base contains the relevant knowledge necessary for understanding, formulating, and solving problems. It includes two basic elements: facts such as the problem situation and the theory of the problem area, and special heuristics or rules that direct the use the knowledge to solve specific problems in a particular domain.
- Inference engine. The inference engine provides a methodology for reasoning about information in the knowledge base and on the blackboard, and for formulating the conclusions.
- User interface. Experts system contain a language processor for friendly, problem - oriented, communication between the user and the computer. This communication can best be carried out in a natural language.

Rules definitions included in knowledge base let precise estimation of a meter signal. Then, depending on analysis, they either result in a transfer to another signal or generation of the appropriate state and message. Decision trees implementation allows for precise fault detection - generated message describes object's state or its part, reason for generated message and a place of occurrence. It lets quick evaluation of a reason for the state and a reaction as a consequence take place. Expert system can change an existing situation - make it better or worse - by gaining a new signal that corresponds with another rule.

4 Database Architecture

Designing of a diagnostic system is determined by needs of the system's users. Operations personnel, process engineers and management need to examine not only what is taking place while current processes in real time but also what happened in the past. These users need access to the actual process data and the summary data, to analyze process operations and make effective decisions.

In addition, many on - line applications such as supervisory control, advanced control, statistical process control, data reconciliation, expert systems, neural networks and other manufacturing execution system functions require historical data. All those systems need to process and analyze actual data and the summarized data over a long period of time. This set of requirements led to the development of the process of the historical data storage.

The historical data storage may record summary data for use in accounting and planning by other system applications. But the most important problem is the acquisition and long - term maintenance of the large volumes of actual process data generated by most manufacturing facilities.

While gathering and managing on - line data four factors are important [9]:

- the data storage efficiency,
- the precision of the captured data,
- the speed at which the data can be written to the database,
- the speed of access to the data.

The relatively inexpensive hard drives available today make storage efficiency appear less important, however they still require a commitment of precious staff resources for data management and hardware maintenance.

There are several basic approaches to data storage and management [4]:

- traditional data storage techniques developed in the first process computer applications,
- relational database management systems,
- multiple databases,
- tables' partitioning,
- process historical data using data compression.

Each of these approaches has its own advantages and disadvantages which need considering while database designing.

4.1 Traditional Data Storage Techniques

Traditional data storage techniques originated in the early days of process monitoring where raw analog data values were stored upon each acquisition cycle and placed in flat - file databases. Storage technologies relied on slow - speed, high - volume media like magnetic tape or high - speed lowvolume media like magnetic core memory or dynamic memory. Acquisition of the process data was limited by the access speed of the equipment and the resolution of the signal conditioning circuitry. That is why the acquisition and storage of large amounts of data was often technically limited and expensive to maintain.

Several approaches were developed to deal with this situation. The simplest was to acquire the data less often, meaning a slower sample rate (for example once every minute or every several minutes). As a result less total data was acquired and more time was recorded per unit capacity of storage media. But it meant accuracy sacrificing because not all the data was recorded. Faster data acquisition combined with storage of only the calculated averages also is a way to attempt to represent the actual data accurately yet still avoid storing each piece of raw data.

Both of the above approaches may miss recording key data values or misrepresent significant changes in the data. When the final user wants to review the recorded data, it is often impossible to tell precisely what occurred because of missing data or because averaged data either exaggerates or mitigates what actually happened. Neither method can reproduce the real actual raw data well, especially if the data is highly variable, as is often the case in industrial processes.

A third approach is to filter or exception report the data. It means recording only those data values that exceed a data tolerance level expressed as a percent of range or a fixed tolerance. This method gives better results in the actual raw data.

4.2 *Relational Database Management Systems*

Most database today are based on a relational model. There are many popular suppliers of Relational Database Management Systems (RDBMS) software, including Oracle, Informix, Sybase, IBM and Microsoft.

The relational model, based on the work of E. F. Codd, has three major components: structures, operations, and integrity rules. Structures are the objects that define how to store the data in a database. Relational database systems have a logical and physical structure. The physical structure is determined by the operating system files. The logical structure is determined by the table spaces and the database's schema objects, which include the tables, views, indices, clusters, sequences, and stored procedures. Typically, one or more data files on the physical disks of the computer are explicitly created for each table to physically store the table's data. The logical structure of the RDBMS defines names, rows, and columns for each table. The columns generally have data types such as character, variable length character, number, date and time, long, raw, long raw, row ID, etc. Once the table name and columns have been defined, data (rows) can be inserted into the database. The table's rows can then be queried, deleted, or updated. Data are retrieved from the tables using Structured Query Language (SQL). SQL statements query the database by searching the table directories for the logical columns and rows that contain the specific data of interest and then returning the results. A procedure is a set of SQL statements grouped together as an executable unit to perform a specific task, for example, to retrieve a set of data with common characteristics. Stored procedures are predefined sets of SQL statements that can be executed together.

The relational database management systems have several important disadvantages compared to modem data compression - based systems:

- The relational database management systems carry more overhead due to their multiple row and table record structures. The use of indices, clusters, and procedures that may slow down the performance considerably.
- All relational database management systems' records are equally important - they are not optimized for the time.

- The relational database management systems have no inherent data compression method. They are usually combined with averaging techniques, which may result in data loss and inaccurate data.
- The speed of writing to relational database management systems is quite slow compared to data compression based systems. The relational database management systems have very high transactions per second, but this refers to actions taken on the data already put in the database, not to the speed at which it is written to the database nor the speed of data retrieval.
- Data statistics are not automatically calculated by the relational database management systems. SQL mathematics is limited to sums, minimums, maximums, and averages.

In order to minimize storage space the data capture rate must be slowed down or only averages should be stored. Slowing the data capture rate or only storing averages all result in discarding information about the manufacturing process. The data's variation, especially fast "spikes", are not captured, and so the corresponding ability to troubleshoot and optimize the manufacturing process is lost.

4.3 Multiple Databases

Data in a database might be put all in a single large database, or placed in multiple smaller databases. Generally, if the data is highly similar and used the same way, then it is better to use one large database rather than several smaller databases. Besides it seems it would be more efficient to use one database instead of creating separate smaller databases because of taking advantage of SQL Server's ability to cache data, such as the data used in lookup tables. The same applies to common stored procedures used in each database. This would present an inefficient use of SQL Server's memory and could potentially reduce performance [1].

4.4 Tables' Partitioning

Data partitioning involves splitting out the rows of a table into multiple tables (horizontal partitioning) or splitting out the columns of a table into multiple tables (vertical partitioning).

If the designed database is supposed to be very large, horizontally partitioning of large tables should be taken under consideration. Horizontal partitioning divides a single table into multiple tables, creating many smaller tables instead of a single, large table. The advantage of this is that is generally is much faster to query a single small table than a single large table.

For example, if a table is expected to add 10 million rows a year, after five years it will contain 50 million rows. In most cases, it may be found that most queries on the table will be for data from a single year. Then partitioning the table into a separate table for each year of transactions will significantly reduce the overhead of the most common of queries.

A table can be horizontally partitioned by:

- time dimension: An important retention period, such as week or month can be used or the data can be partitioned by the age of the data (e.g. if the analysis is usually done on last month's data the table could be partitioned into monthly segments),
- a dimension other than time: If queries usually run on a grouping of data (e.g., each branch tends to query on its own data) and the dimension structure is not likely to change then partition the table on this dimension,
- on table size: If a dimension cannot be used, partition the table by a predefined size. If this method is used, metadata must be created to identify what is contained in each partition.

Row splitting creates a one - to - one relationship between the partitions. Normalization creates one - to - many relationships between the partitions. One - to - many relationships can lead to large table joins which would decrease, not improve, performance.

Taking advantage of horizontal partitioning requires early planning as the various queries developed for the application need to be designed to take advantage of the database's design. Vertical partitioning optimizes performance since the early days of relational databases.

A fact table can be vertically partitioned by:

- normalizing the table,
- moving seldom used columns from a highly - used table to another table, this is sometimes called row splitting.

Combined with a clustering algorithm, affinity determines a reasonable assignment of attributes to vertical fragments. Attribute affinity identifies clusters by collecting statistics about the attribute usage by queries, and can therefore scale to large workloads. Its disadvantage is that it is decoupled from the system's optimizer and the query execution engine, and thus human intervention is eventually required to validate the quality of the recommended partitioned designs.

4.5 The Diagnostic System's Database Architecture

The acquisition of the data concerning object's state results in data storage for the further display or better diagnosis of the object's state in the future. Proper logical database design results in proper physical database design, and generally results in good database performance. Both the logical and physical design must be well - considered to get good performance out of the database. If the logical design is not right before beginning the development of the application, it is too late after the application has been implemented to fix it.

The choice of the database management system and the system's application results in some limits in the system's architecture. It also influences in the ways and possibilities of the diagnostic system's management and administration.

Depending on a kind of measurements, five different files are generated:

- temperature file,
- tap changer power consumption file,
- file containing data on extraction of moisture and hydrogen dissolved in oil,
- overvoltage file,
- other measurements (including binary values).

Every 2 seconds over 150 signals' values are put into those files. The system process automatically loads meter data. Initial signal estimation takes place while reading in files in order to exclude incorrect or redundant signals. Considering the time restrictions it was necessary to solve the problem of the files' loading with the assumed system functionality.

Data storing method has been constrained by the way meter signals are saved into files as well as by the presentation of the off - line data. It is also the most optimum method for object diagnosing and data presentation. The project of tables and field types ensures system efficiency despite large amount of data being stored.

The diagnostic system is based on relational database management system. Besides it uses tables' partitioning. Tables are horizontally partitioned by time dimension - a single year is placed in one table. It allows most of the queries to be much faster because the queries that use all the data are very rare.

Microsoft SQL Server is the core of the monitoring system [27, 28, 29]. All data loaded from text files generated by meter circuits are put in MS SQL Server database. Nevertheless the structure of the diagnosing system enables access to stored data using any database platform. The correctness of the monitoring system has been checked using PostgeSQL database platform as well. The schema of a database is shown in figure 4. The table, named tblObject, is the central and fundamental part of diagram. It contains basic data concerning diagnosed power transformers (object's ID, its name, number and the year of the production). All the other objects' data are stored in separate tables that are joined together using different kind of relations. Considering characteristics of data to be stored, the off - line and on - line signals have been put into separate tables. In addition rating data, which stay changeless while exploitation of transformer, has been kept apart. All the data enable the precise identification of the objects for the technical stuff.

The object has been divided into four elementary modules:

- active part
- oil,
- cooling unit,
- tap changer.

They are called Locations within a database. Each of these locations refers to particular signals, although some of the signals might refer to more than one location. Object division into modules allows diagnosing and localizing the malfunction more accurately. There is also the possibility to extend the system by adding further modules for the storage of the information coming from the new meter circuits or new objects if necessary.

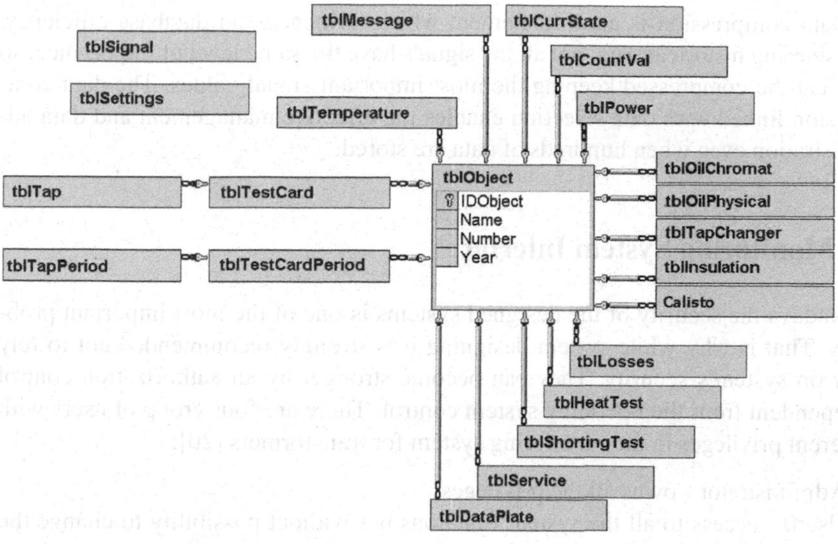
tblMessage
tblCurrState
tblSignal
tblCountVal
tblSettings
tblTemperature
tblPower
tblTap
tblTestCard
tblObject
IDObject
Name
Number
Year
tblOilChromat
tblOilPhysical
tblTapChanger
tblInsulation
Calisto
tblTapPeriod
tblTestCardPeriod
tblLosses
tblHeatTest
tblShortingTest
tblService
tblDataPlate

Fig. 4 The schema of the database

There are four technical states, which refer to each of the modules and a transformer as a whole.

- 0 - normal
- 1 - normal
- 2 - warning
- 3 - alarm
- 4 - emergency

They are called States in the database and refer directly to Messages table and Current State table. Considering expert knowledge the rules' database has been created. Rules definitions included in knowledge base decide on object's state or its part, reason for generated message and a place of occurrence. Most of meter signals are subject to analysis and diagnosis while they are being loaded.

All the rules used by an expert system are gathered in appropriate tables to enable updates of the system when the number of sensing elements increase. The procedures concerned meter signals are either SQL statements or programming codes.

While database administration, it is necessary to archive historical data. The final user can reach historical data by choosing appropriate option from user interface menu. Sometimes it demand additional actions to be taken, especially if the information is very old - tables containing signals' values are horizontally partitioned and compressed.

The appropriate way of data archive lets database to work more efficient and reduces database increments. It is very important for those databases which store lots of information. As it was written before, the database management system influences data storage and systems administration which influences the system efficiency and data management as a consequence.

Data compression is another element which influences on database efficiency. Considering historical data, not all the signals have the same level of importance, so they can be compressed keeping the most important signal values. The data compression linked with data selection enables the effective management and data administration even when hundreds of data are stored.

5 Monitoring System Interface

Nowadays the security of the designed systems is one of the most important problems. That is why while system designing it is strongly recommended not to rely only on system's security. They can become stronger by an authorization control independent from the operating system control. There are four group of users with different privileges in the monitoring system for transformers [20]:

- Administrator - owns all the privileges,
- User0 - access to all the system functions but without possibility to change the Settings,
- User1 - like the User0 without possibility to edit the off - line data,
- User2 - just observes the on - line data.

Moreover the remote user of the diagnostic system must log on by entering remote user name and a password.

The diagnostic system for large power transformers security consists of three layers:

- Windows Server security - the user has to exist in the system,
- remote access security - the user has to belong to the group of remote users,
- diagnostic system security - the user has to know the password for one of the groups described above.

The data concerned users are stored within the database in a separate table specially encoded.

The final user of the monitoring system can obtain a clear information on existing situation based on analysis and estimation. Similarly the return to normal state is immediately registered. Except for numeric data, a state message and a transformer's state change are generated an inserted into the database.

For better visualization there are four colors that correspond to particular states:

- green for normal state,
- yellow for warning state,
- orange for alarm state,
- red for emergency state.

If there is a situation when several signals change their state for worse, signals that make the situation worse are top priorities. Meter signals' diagnosis is proceeded the same way as binary signals' estimation. Moreover, there is a possibility to change acceptable limits' criteria for meter signals (only system administrator).

Some of data (off - line), entered by a user in irregular time intervals, is analyzed and diagnosed as well. In case of off - line data, change for the better is possible when new data, which fulfill exploitation criteria, is entered.

Meter signals influence additional numeric characteristics, which are subject to continuous estimation and analysis regardless of reading files frequency. It refers to transformer aging inter alia.

Basing on a part of signals transformer's state, forecast can be made as well as future trends in signals values' changes can be analyzed [23].

There is a loading time calculator for overload conditions. Overload has been divided into two separate areas. One of them allows counting acceptable loading time in a long - term loading conditions (which means multiplication factor $k \leq 1.3$ for rated current IDN). The second area allows counting acceptable loading time in a short - term loading conditions (which means multiplication factor $1.3 < k \leq 1.5$ for rated current IDN) [20, 3].

Future trend analysis refers to extraction of moisture and hydrogen dissolved in oil. It is based on values stored for a specific time interval (three months, a month, a week, twenty - four hours). These values, extrapolated with exponential function, allow trend observing for a month forward.

6 Data Compression Techniques

The development of data compression techniques that record points at unequally spaced time intervals stemmed from three factors:

- the need to access ever increasing amounts of data,
- the advances in minicomputer technology provided the processing power to enable collection of the data,
- the recognition that there are long periods of operation in which variables are either constant or moving in a predictable path inspired the solution.

Data compression is a storage methodology that uses the predicted behavior of a process parameter to reduce the actual raw data values to a data subset, which can then be later expanded and interpolated (decompressed) to produce a representation of the original data accurate to within the data tolerance level specified. Databases based on these techniques can store data over long time periods using the minimum storage space.

Database systems based on data compression can store data over long time periods using the minimum storage space. Accuracy is ensured because the compression algorithm selects only those points for storage that are necessary to reproduce the actual data within the tolerance specified. No further data reduction or file compression techniques are necessary to reduce the storage space. Accurate statistics on the data (averages, standard deviations, accumulated totals, minimums, maximums, etc.) within a specified time can be readily generated using proven mathematical techniques for time series data.

There are many different algorithms of compression. Three following methods used in the monitoring system are described below:

- Run Length Encoding algorithm,
- LZW algorithm,
- Huffman algorithm.

6.1 Run Length Encoding Algorithm Bases

Run - length encoding "RLE" is a form of data compression in which runs of data (that means sequences in which the same data value occurs in many consecutive data elements) are stored as a single data value and count, rather than as the original run. This is most useful on data that contains many such runs.

The general algorithm of RLE replaces any sequence of identical symbols by a counter identifying the number of repetitions and the particular symbol [10].

6.2 Signal Compression Using Huffman Algorithm

The Huffman algorithm is a dynamical programming algorithm that constructs a binary tree that minimizes the average bit rate. This tree is called an optimal prefix code tree [10].

The binary tree of nodes can be stored in a regular array, the size of which depends on the number of symbols. A node can be either a leaf node or an internal node. Initially, all nodes are leaf nodes, which contain the symbol itself, the weight (frequency of appearance) of the symbol and optionally, a link to a parent node which makes it easy to read the code in reverse - starting from a leaf node. Internal nodes contain symbol weight, links to two child nodes and the optional link to a parent node. As a common convention, bit '0' represents following the left child and bit '1' represents following the right child. A finished tree has n leaf nodes and n - 1 internal nodes.

A linear - time method to create a Huffman tree is to use two queues:

- the initial weights with pointers to the associated leaves are put in the first queue,
- the combined weights with pointers to the trees are put in the back of the second queue.

This assures that the lowest weight is always kept at the front of one of the two queues.

Huffman coding is optimal when the probability of each input symbol is a negative power of two.

6.3 Signal Compression Using LZW Algorithm

Lempel - Ziv - Welch "LZW" is a universal lossless data compression algorithm created by Abraham Lempel, Jacob Ziv, and Terry Welch [10]. The algorithm of the compression builds a string translation table from the text being compressed. The string translation table maps fixed - length codes to strings (the codes are usually 12 - bit length). The string table is initialized with all single - character strings (256 entries in the case of 8 - bit characters). As the compressing algorithm character examines the text, it stores every unique two - character string into the table as a code concatenation, with the code mapping to the corresponding first character. When each two character string is stored, the first character is sent to the output. Whenever a previously encountered string is read from the input, the longest such a previously encountered string is determined, and then the code for this string concatenated with the extension character (which is the next character in the input) is stored in the table. The code for this longest previously encountered string is outputted and the extension character is used as the beginning of the next string.

The decompressing algorithm requires the compressed text as an input. It can build an identical string table from the compressed text as it is recreating the original text. However, an abnormal case shows up whenever the sequence character - string - character - string - character (with the same character for each character and string for each string) is encountered in the input and character - string is already stored in the string table. When the decompressing algorithm reads the code for character - string - character in the input, it cannot resolve it because it has not yet stored this code in its table. This special case can be dealt with because the decompressing algorithm knows that the extension character is the previously encountered character.

6.4 Data Compression Used in the Diagnostic System

Data may be compressed using one of the following methods [7]:

- before inserting samples it into database,
- by table compression using internal database compression tools,
- by compressing database files.

In the monitoring system described in this paper input samples are grouped into blocks, compressed and inserted into another database.

The compression algorithm separates the input data to enable optimal compression using Lempel - Ziv - Welch (LZW) or Huffman coding algorithms. In the large power transformer monitoring system RLE algorithm is used to split the input data into LZW - type and Huffman - type. The RLE algorithm encodes values and run lengths of the input data. The input data values can be optimally compressed using Huffman coding and run length values can be optimally compressed using LZW algorithm [6].

The compression is performed in the following steps:

- the input data are divided into blocks (e.g. one month),
- the input data are transformed using RLE algorithm,
- the run length output is compressed using LZW algorithm
- the data value output is compressed using Huffman algorithm.

All uncompressed data are stored in one database while compressed data are stored in another database (fig.5).

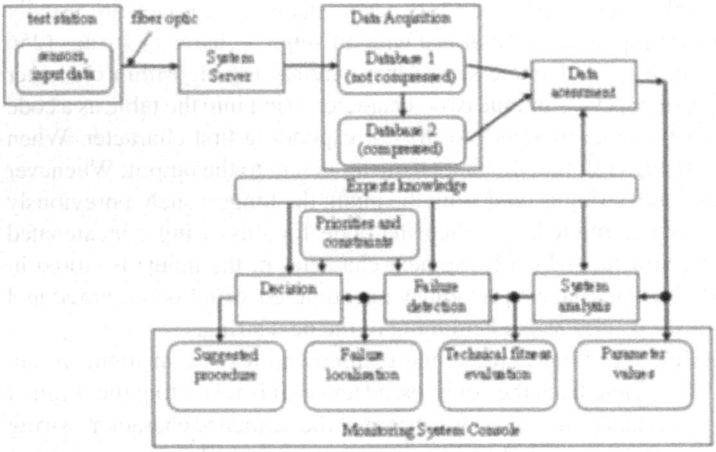
test station
fiber optic
sensors, input data
System Server
Data Acquisition
Database 1 (not compressed)
Database 2 (compressed)
Data assessment
Experts knowledge
Priorities and constraints
Decision
Failure detection
System analysis
Suggested procedure
Failure localization
Technical fitness evaluation
Parameter values
Monitoring System Console

Fig. 5 Logic structure of the power transformer system

To perform decompression, data are transformed in reverse direction to the compression method. To avoid the time consuming decompression in the structure mentioned above it is possible to obtain the coarse values of the compressed data. It may be used when long time periods are analyzed. In such cases only Huffman - based decompression should be performed and run length information is neglected. It causes time scale inconsistency but the signal values (which is the most important) are perfectly reconstructed.

7 Conclusions

The problem of diagnosing the state of any device is of increasing importance because of its fundamental part in the whole process of supervising work of this device. Development of the computer systems and meter circuits enable very precise and reliable condition diagnosis. Moreover on - line diagnostic systems bring economical benefits because of the prevention of major failures costs for outages, repair, and associated damages.

The process of device's diagnosis needs capturing, managing and providing access to tens of thousands of related data values that are acquired at rates from milliseconds to a few minutes. Data are stored on - line for couples of years and increase every minute. The process of storing data based on the compression methods is specifically designed to deal such situations using minimum storage space. Data compression methods can provide high efficiency and accuracy. On the basis of the discussed diagnostic system for the large power transformers it can be estimated that usage of the compression techniques saves even up to 90 percents of the capacity of the storage space and still run relatively narrow histogram characteristics. The signal values may be also perfectly reconstructed and have no drawbacks generated by averaging techniques.

References

1. Asaro, T.: Intelligent Data Archiving. The Enterprise Strategy Group, Inc (2007)
2. Bugajny K., Kaźmierski M., Kersz I., Pinkiewicz I., Wosiak A., Olech W., Rzeczkowski A., Szymański Z.: System monitoringu transformatorów sieciowych weksploatacji, oprogramowanie systemowe
3. Cichosz, P.: Systemy uczace sie, Wydawnictwa Naukowo–Techniczne, Warszawa (2000)
4. Keene, W.N.: Large Database Alternatives. NEON Enterprise Software, Inc. (2007)
5. Korbicz, J., Kościelny, J.M., Kowalczuk, Z., Cholewa, W.: Diagnostyka procesów, WNT Warszawa (2002)
6. Lipinski, P., Puchala, D., Wosiak, A., Byczkowska–Lipinska, L.: Transformer Monitoring System Taking Advantage of Hybrid Wavelet Fourier Transform. In: ISEF 2007—XIII International Symposium on Electromagnetic Fields in Mechatronics, Electrical and Electronic Engineering Prague, Czech Republic, September 2007, s 126–129 (2007)
7. Lipiński, P.: A New Adaptive Wavlet Transform. In: Proceedings of the 11th International Conference on System Modelling Control, Warszawa 2005, s 179–188 (2005)
8. Mulawka, J.: Systemy ekspertowe. WNT Warszawa (1996)
9. Mullins, C.S.: Database Archiving: Managing Data for Long Retention Periods, NEON Enterprise Software, Inc. (2007)
10. Sayood, K.: Kompresja danych. Wprowadzenie, Wydawnictwo RM, Warszawa (2002)
11. Baum J., Dornemann K., Stirl T., Tenbohlen S.: Enhanced Diagnosis of Power Transformers using On—and Off–line Methods: Results, Examples and Future Trends, Cigre, paper 12–204, Paris, 2000
12. Liebfried, T., Knorr, W., Kosmata, A., Sundermann, U., Viereck, K., Dohnal, D., Breitenbauch, B.: On–line Monitoring of Power Transformers–Trends. In: New Developments and Firs Experiences, Cigre, 12–211, Paris (1998)
13. Pettersson, L., Fantana, N.L., Sundermann, U.: Life Assessment: Ranking of Power Transformers Using Condition Based Evaluation. A New Approach. Cigre 12–204, Paris (1998)
14. Provanzana, J.H., Gattens, P.R., Hagman, W.H., Moore, H.R., Harley, J.W., Triner, J.E.: Transformer Condition Monitoring—Realizing an Integrated Adaptive Analysis System, Cigre 12 - 105, Paris (1992)
15. Kavamura, T., Fushimi, Y., Shimano, T., Amano, N., Ebisawa, Y., Hosokawa, N.: Improvement in Maintenance and Inspection and Pursuit of Economical Effectiveness of Transformers in Japan, Cigre 12–107, Paris (2002)
16. Boss, P., Lorin, P., Viscardi, A., Harley, J.W., Isecke, J.: Economical aspects and practical experiences of power transformer on - line monitoring, Cigre 12–202, Paris (2000)
17. Aschwanden, T., Hassig, M., Fuhr, J., Lorin, P., Houhanessian, V., Zaengl, W., Schenk, A., Zweiacker, P., Piras, A., Dutoid, J.: Development and application of new condition assessment methods for power transformers, Cigre 12–207, Paris (1998)
18. Carbonell, J.G., Michalski, R.S., Mitchell, T.M.: An overview of machine learning. Machine learning: An Artificial Intelligence Approach, Vol. 1, Tioga (1983)
19. Byczkowska–Lipińska, L., Wosiak, A.: Projekt inteligentnego systemu diagnostyki wibroakustycznej, Materialy konferencyjne. In: VI Krajowa Konferencja Naukowo–Techniczna DPP 2003 (2003)
20. Byczkowska - Lipińska L., Wosiak A.: Architecture of Monitoring System For Large Power Transformers. In: Techniczna Elektrodynamika, Kijów 2006, no. 6, str.: 32–35 (2006) ISSN 0204–3599

21. Byczkowska–Lipińska L., Wosiak A.: Zastosowanie Systemu Informatycznego w Diagnostyce Transformatorów. In: XVI Sympozjum PTZE, Zastosowania Elektromagnetyzmu w Nowoczesnych Technikach i Informatyce, Wisa 25–27 wrzenia 2006, str. 73–75 (2006) ISSN 1233–3336
22. Wosiak, A.: System monitorowania pracy transformatora energetycznego. In: V Sympozjum Modelowanie i Symulacja Komputerowa w Technice, Wyzsza Szkola Informatyki, Lodz 2006, str. 133–136 (2006)
23. Kaźmierski, M., Kersz, I., Wosiak, A.: System monitoringu stanu transformatorów w ukadzie "on - line", realizacja praktyczna. Konferencja Zarzadzanie Eksploatacja Transformatorów, Wisa Jawornik 26–28 kwietnia 2006, str. 121–134 (2006)
24. Byczkowska - Lipińska L., Wosiak A.: Monitoring System of the State of Electric Devices. In: Proceeding of the International Conference TCSET, Modern Problems of Ratio Engeneering Telecomminications and Copmuter Science, Lviv - Slavske, Ukraine, February 28–March 4 2006, str. 31–33 (2006)
25. Kaźmierski, M., Kersz, I., Wosiak, A.: System monitoringu stanu transformatorów energetycznych, wstepne doświadczenia eksploatacyjne, Przeglad Elektrotechniczny Konferencje 2006, nr 1, ss. 121–124 (2006)
26. Byczkowska-Lipińska, L., Wosiak, A.: System diagnostyczny do oceny stanu pracy transformatora energetycznego, Przeglad Elektrotechniczny nr 12/2006 (2006)
27. Gunderloy, M., Chipman, M.: SQL Server 7, Mikom, Warszawa (1999)
28. Gallagher, S.: Microsoft SQL Server 7. Ksiega Experta, Helion, Warszawa (2000)
29. Riordan, R.M.: Programowanie Microsoft SQL Server 2000 krok po kroku, Read–Me, Warszawa (2001)
30. Brachman, R., Levesque, H.: Readings in Knowledge Representation. Morgan Kaufmann, Los Altos (1985)

Subject Index